Manual of Water Supply Practices

M64

Gas Transfer Applications in Water: Addition and Removal

American Water Works Association

Manual of Water Supply Practices—M64

Gas Transfer Applications in Water: Addition and Removal

If you find errors in this manual, please email books@awwa.org. Possible errata will be posted at www.awwa.org/M64

Managing Editor – Book Products: Melissa Valentine
Project Manager/Technical Editor: Suzanne Snyder
Cover Design/Technical Illustrations: Michael Labruyere
Manager – Publishing Operations: Gillian Wink
Production: Innodata
Senior Specialist – Manuals: Willadee Hitchcock

Library of Congress Cataloging-in-Publication Data
Names: Baldwin, Jennifer (Water purification), author. | American Water Works Association.
Title: M64 - gas transfer applications in water : addition and removal / by Jennifer Baldwin.
Other titles: Gas transfer applications in water: addition and removal | AWWA manual ; M64.
Description: First edition. | Denver, CO : American Water Works Association, [2021] | Series: Manual of water supply practices ; M64 | Includes bibliographical references and index. | Summary: "This manual is designed to provide operators and engineering staff with an understanding of aeration and air stripping systems, with a comprehensive look at different technologies and applications of gas addition and gas removal"-- Provided by publisher.
Identifiers: LCCN 2020036584 (print) | LCCN 2020036585 (ebook) | ISBN 9781647170264 (paperback) | ISBN 9781613005668 (adobe pdf)
Subjects: LCSH: Water--Aeration--Handbooks, manuals, etc. | Water--Purification--Handbooks, manuals, etc. | Waterworks--United States--Handbooks, manuals, etc.
Classification: LCC TD458 .B352 2021 (print) | LCC TD458 (ebook) | DDC 628.1/65--dc23
LC record available at https://lccn.loc.gov/2020036584
LC ebook record available at https://lccn.loc.gov/2020036585
Printed in the United States of America

ISBN 978-1-64717-026-4 ISBN, electronic 978-1-61300-566-8

DOI https://doi.org/10.12999/AWWA.M64ed1

American Water Works Association

American Water Works Association
6666 West Quincy Avenue
Denver, CO 80235-3098
awwa.org

Contents

Figures

This page intentionally blank.

Tables

This page intentionally blank.

Preface

A group of volunteers from the Inorganics Committee of the American Water Works Association (AWWA) prepared this manual of practice. The need for a manual on aeration and air stripping was brought to the attention of the committee at the 2010 Inorganic Contaminants Workshop. It was discussed at the workshop that a standalone manual on aeration and air stripping did not exist. The goal of this first edition of M64 is to provide the reader with a general understanding of the principles of aeration and air stripping, providing knowledge regarding the applications of aeration and air stripping, and providing operators guidance on the operation and maintenance of aeration and air stripping equipment, including case studies. The manual includes Henry's law constants for several contaminants of interest in drinking water applications. We encourage interested readers to pursue additional information on aeration and air stripping topics through other AWWA resources.

The contents of this Manual include the importance of gas transfer operations in water treatment, including a definition of aeration (Chapter 1), the physical mechanism of gas transfer, including Henry's law constants and information regarding blower selection (Chapter 2), the principles and applications for gas addition (Chapter 3), the principles and applications for gas removal (Chapter 4), and the installation, operation, and maintenance of aeration and air stripping equipment (Chapter 5). Chapter 2 contains a summary table of Henry's law constants, and Chapter 5 includes case studies of existing system operation and maintenance. The index can be helpful in finding information on a subject of interest presented in different context throughout the manual.

This manual is designed to provide operators and engineering staff with an understanding of aeration and air stripping systems, with a comprehensive look at different technologies and applications of gas addition and gas removal. The Inorganics Committee is hopeful that M64 will meet industry needs and will be a useful resource for a broad audience. As this is the first edition of AWWA Manual M64, *Gas Transfer Applications in Water: Addition and Removal*, the Inorganics Committee and AWWA welcome comments and suggestions for improving future editions of this manual. Please contact the AWWA Water Quality Engineer, 6666 West Quincy Avenue, Denver, CO 80235-3098 to provide feedback on the contents of this manual.

Jennifer Baldwin, Ph.D., P.E.
Chair, M64 Committee

This page intentionally blank.

Acknowledgments

The AWWA Technical & Educational Council, the Water Quality & Technology Division, and the Inorganics committee gratefully acknowledge the contributions made by those volunteers who drafted, edited, and provided the significant and critical commentary essential to developing M64. The Steering Committee members dedicated many hours in the final stages of preparation of this edition to ensure the overall technical quality, consistency, and accuracy of the manual.

Steering Committee Members

Jennifer Baldwin, Chair, Jacobs, Knoxville, Tenn.
Thomas Gillogly, Carollo Engineers Inc., Wilton Manors, Fla.
Patrick Gurian, Drexel University, Philadelphia, Pa.
Jerry Lowry, Lowry Engineering, Inc., Blue Hill, Maine
Jennifer Santini, AWWA Staff, Denver, Colo.

Contributors to the 1st edition

Jennifer Baldwin, Chair, CH2M, Knoxville, Tenn.
Marnie Bell, Arcadis, Buffalo, N.Y.
Frederick Bloetscher, Public Utility Management and Planning Services, Inc., Hollywood, Fla.
Nicole Blute, Hazen and Sawyer, Los Angeles, Calif.
William Byers, CH2M, Corvallis, Ore.
Patrick Cole, H2M Architects + Engineers, Howell, N.J.
Catherine DiPietro, Aquarius Engineering, Pownal, Maine
Joseph Drago, Kennedy/Jenks Consultants, San Francisco, Calif.
Thomas Gillogly, Carollo Engineers Inc., Wilton Manors, Fla.
Angelica Gregory, Carollo Engineers Inc., Coral Springs, Fla.
Pierre Kwan, HDR Engineering, Inc., Seattle, Wash.
Milton Larsen, Kennedy/Jenks Consultants, Federal Way, Wash.
Jerry Lowry, Lowry Engineering, Inc., Blue Hill, Maine
Lee O'Dell, CH2M, Portland, Ore.
Michael Schock, Environmental Protection Agency, Cincinnati, Ohio
Samuel Silva, California Water Service Co., Menlo Park, Calif.

This page intentionally blank.

Chapter 1

Importance of Gas Transfer Operations in Water Treatment

OVERVIEW

Aeration is a well-established and relatively common process that has been used for producing drinking water for many centuries. The industry definition of aeration per the AWWA Water Dictionary is "a gas transfer unit process that allows for the absorption of gas (frequently oxygen or disinfectant) by water, the air stripping of volatile compounds, and the precipitation of inorganic contaminants."

This short definition covers a broad range of processes that involves gas transfer into and out of water. This chapter is intended to provide an overview of the aeration process, how it was developed and has been refined to its current design, why it has been used, what gases are added into or removed from water, and contaminants it can be used to remove. From this overview and introduction, readers should gain a basic understanding of the aeration process before seeking out the more topic-specific chapters later in the manual to address their individual questions and requirements.

HISTORY AND FUTURE

Aeration has been used since the earliest civilizations as people recognized the benefit of mixing water with air. Those humans found that adding air removed odors and improved the taste of surface and groundwaters. In those earliest aeration applications, the aeration process was initiated by bubbling air through basins full of water, having water cascade across manmade waterfalls, or simply exposing basins of water to windy conditions. All of these ancient uses were to address taste and odor issues before drinking the water, via a combination of transferring oxygen into the water (oxygenation) and the removal of noxious dissolved gases out of the water (air stripping).

Until the 1970s, aeration was still primarily used to address aesthetic issues, such as taste and odors and iron oxidation before filtration. Its use for other water treatment in North America generally started in the late 1970s when the United States Environmental Protection Agency (USEPA) began the regulation of volatile organic compounds (VOCs) as a result of the mandated cleanup of VOC-contaminated sites. The VOC contamination entered groundwater at many of these sites, with subsurface plumes oftentimes extending close to or into the drinking water supplies. Many VOCs are hydrophobic and readily removed using aeration. As a result, numerous "pump-and-treat" aeration systems were installed to remove the contaminants.

Since then, aeration has been identified by USEPA as one of the best available technologies to directly treat many primary drinking water contaminants, including synthetic organic chemicals (SOCs) and radon. In addition, aeration has been identified as a process to indirectly address several other drinking water rules. Examples include the Lead and Copper Rule for the removal of carbon dioxide in membrane-treated waters and to address plumbing corrosion issues associated with low-pH waters and lower total inorganic carbon. Another example is using aeration to remove total trihalomethanes (TTHMs), as the means for utilities to achieve compliance with the Stage 1 and Stage 2 Disinfectants and Disinfection Byproducts Rules. Finally, aeration is often used to remove methane, an unregulated contaminant, from groundwater in order to prevent hazardous conditions from developing due to the flammability of methane.

Looking forward at the time of this writing, USEPA's proposed carcinogenic VOCs (cVOCs) rule and potential reductions in regulatory limits for individual VOCs could lead more utilities to implement aeration. Also, more utilities are facing the dilemma of continuing to meet their communities' water demands without any available high-quality water sources. Meeting the demand means that these utilities may need to reconsider the use of previously ignored or bypassed water sources. These sources are generally of poorer quality and often contain contaminants for which aeration can be a viable treatment solution.

COMPARATIVE BENEFITS AND DISADVANTAGES OF AERATION VERSUS OTHER TREATMENT PROCESSES

The aeration process is a very appealing treatment as it is often inexpensive compared to other processes. Also, it most often uses no chemicals, thereby eliminating the expense of chemical delivery systems, storage tanks, transfer pumps, and piping. The lack of chemicals also eliminates the community and operator exposure risks associated with delivering chemicals, cleaning up chemical spills, and potential chemical over- or under-dosing. This inherent safety makes aeration a more attractive option than caustic soda addition for pH adjustment or hypochlorite/permanganate for iron oxidation (its use for solely manganese oxidation is not as common due to the longer reaction times).

Aeration also has a distinct advantage over activated carbon processes in addressing SOCs and TTHMs. Aeration does not produce a waste residual from the treatment process except in cases where air quality regulations require removal of the contaminant from the vapor phase. In comparison, activated carbon needs to be removed, replaced, and properly landfilled once the carbon's absorptive capacity is exhausted (unless it is possible to regenerate the carbon either thermally or through biodegradation). In addition, aeration requires considerably less energy compared to reverse osmosis membranes.

The aeration system does have some disadvantages associated with it. Aeration systems can be quite large, especially when treatment flowrates are large, the required amount of contaminant removal is great, or if the required treated water discharge concentration is very low, i.e., <1 µg/L. These conditions result in large diameter, tall vessels if packed towers are used. Such towers may conflict with local zoning requirements and neighborhood concerns over visual impacts and the release of contaminants into the air, especially if the installation could be located in a residential area. Large diffuser array basins can be used but the footprint may be prohibitive, and there may be construction difficulties if the basins need to be deep. Once the requisite blowers, air piping, and electrical panels are included, aeration towers or basins often require more land than do some other treatment processes. Aeration systems often operate at atmospheric pressure, so pumping into and out of these systems is often required. The additional pumps add process complexity, capital costs, energy costs, and further increase the space requirements. Chemicals may be required to treat the contaminants stripped from the water (hydrogen sulfide is an example).

Aeration systems generally have a higher capital cost than liquid (caustic soda, sodium hypochlorite, or permanganate) feed systems but may offer lower lifecycle costs and may be preferred because of safety considerations. The costs of aeration systems versus activated carbon absorption are dictated by the relative difference between aeration's pumping and blower energy costs and the labor and materials costs for activated carbon replacement, though aeration has historically been the less expensive of the two.

MANUAL PURPOSE, ORGANIZATION, AND USE

This manual is intended to provide guidance to water quality specialists and engineers with theoretical and practical understanding and concepts for using aeration systems to meet current and potentially future regulatory requirements. To assist the reader, this manual has been organized into distinct, separate chapters relating to specific contaminants or groups of contaminants, aeration system design, and operation and maintenance. As a result, a reader does not necessarily need to read the entire manual to gain the knowledge needed to complete their required tasks. These chapters are:

1. Importance of gas transfer operations in water treatment – this chapter.
2. Physical mechanism of gas transfer – this chapter explains the fundamental chemical equations and reactions in the process followed by the major physical features and considerations for various forms of aeration systems, including weir, cascade, and tray aerators, packed towers, staged diffused bubble systems, and blowers.
3. Gas addition – the processes to add gases to water to improve water quality and/or downstream processes are discussed in this chapter, including air/oxygen, gaseous chlorine, ozone, anhydrous ammonia, sulfur dioxide, and carbon dioxide.
4. Gas removal – also known as "air stripping," this chapter describes how the aeration process can be used to remove various dissolved contaminants, including carbon dioxide, hydrogen sulfide, radon, VOCs, and TTHMs.
5. Installation, operation, and maintenance – this manual's final chapter reviews the issues associated with installing, operating, and maintaining an aeration system.

RESOURCES

As with any design manual and reference material, the authors have made their best effort to provide a comprehensive source of knowledge to the reader, but it is the readers' responsibility to use their professional judgment to properly apply the information provided for their specific requirements. The authors encourage readers to independently verify the information contained herein with other, and potentially more current, sources of information. In addition, readers should review other design guidance manuals. A partial list of other manuals with aeration information includes:

American Water Works Association and American Society of Civil Engineers. 2011 (6th ed.). *Water Quality and Treatment: A Handbook on Drinking Water* (J.K. Edzwald, editor. New York: McGraw-Hill.)

American Water Works Association and American Society of Civil Engineers. 2005 (4th ed.). *Water Treatment Plant Design.* New York: McGraw-Hill.

Crittenden, J.C.; Trussel, R.R.; Hand, D.W.; Howe, K.J.; and Tchobanoglous, G. 2012 (3rd ed.). *MWH's Water Treatment: Principles and Design.* New York: John Wiley & Sons.

Great Lakes – Upper Mississippi River Board of State and Provincial Public Health and Environmental Managers. 2012. *Recommended Standards for Water Works ("Ten States Standards"),* Health Research Inc.

HDR Engineering, Inc. 2001 (2nd ed.). *Handbook of Public Water Systems.* New York: John Wiley & Sons.

In addition, a good source of drinking water-related research for specific contaminants is the Water Research Foundation (www.waterrf.org).

Chapter 2

Physical Mechanism of Gas Transfer

INTRODUCTION

This chapter describes (1) the basic mechanisms of gas transfer into or out of drinking water and (2) the equipment that is designed to carry out the gas transfer. The physical mechanism of gas transfer plays a critical role in water treatment techniques used for contaminant removal and chemical conditioning. Gas transfer mechanisms can be categorized as either (1) air stripping (desorption) or (2) aeration (gas absorption). Air stripping is the transfer of a gas or volatile organic compound from water to air, whereas aeration is the transfer of a gas to water. Air stripping has been used for the removal of carbon dioxide, radon, nitrogen trichloride, oxygen gas, hydrogen sulfide, methane, and many volatile organic compounds (VOCs) from water. Common aeration practices include the addition of carbon dioxide, oxygen, ozone, and chlorine. A full spectrum of water treatment applications of air stripping and aeration are summarized in Table 2-1.

Design of gas transfer equipment requires a basic understanding of gas-water equilibrium described by Henry's law, and the kinetics of mass transfer across the water-gas interface. Gas transfer equipment is designed to optimize the mass transfer for a particular compound while balancing capital and operating costs and footprint requirements.

EQUILIBRIUM AND HENRY'S LAW

In gas transfer, air and water are brought into contact with each other to take advantage of the relative solubility of the gas in water and air. Equilibrium describes the final state of solubility in each phase after the mass rate of transfer between water and air are equal. A system not in equilibrium will always move toward equilibrium. The rate at which the

Table 2-1 Water treatment air stripping and aeration applications

Phase Transfer Type/Compound	Water Treatment Goals
Gas absorption	
O_2 addition	• Oxidation of inorganics, lake destratification
O_3 addition	• Oxidation of organics, disinfection, and color removal
Cl_2 addition	• Oxidation of inorganics and disinfection
ClO_2 addition	• Disinfection
CO_2 addition	• pH control
SO_2 addition	• Dechlorination
NH_3 addition	• Chloramine formation for disinfection
Air stripping	
CO_2 stripping	• Corrosion control
O_2 stripping	• Corrosion control
H_2S stripping	• Odor control
NH_3 stripping	• Nutrient removal
VOC stripping	• Taste and odor removal, VOC removal

Source: Montgomery 1985

liquid-gas system approaches equilibrium is a function of how far the system is from equilibrium. Equilibrium in aeration applications is not typically reached because it is usually not cost-effective to add or remove more than is required by the application.

For gas transfer applications in drinking water, the equilibrium partitioning between water and air can be described by Henry's law:

$$P_a = K_{H,a}\, x_a \qquad \text{(Eq 2-1)}$$

At equilibrium, the partial pressure or concentration of a compound in the gas phase (P_a) is proportional to the mole fraction of the compound in the liquid phase (x_a). The proportionality constant, $K_{H,a}$, is also known as the Henry's law constant. The equation is valid for dissolved organic and inorganic compounds that are able to move between the water phase and air phase in drinking water up to near the solubility of the compound. Typical concentrations of gaseous compounds in drinking water solutions are far below levels that would influence the Henry's law constant. Further, drinking water is generally of such high quality that the ionic strength or surfactants are not factors that will alter the Henry's law constant.

The value of the Henry's law constant provides insight into the relative ease of a compound to move from the water phase to the air phase or vice versa. Air stripping processes are more effective for compounds with higher Henry's law constants, and aeration processes are more effective for compounds with lower Henry's law constants.

The Henry's law constant describes only the equilibrium state and does not describe the rate of transfer. Further, it must be pointed out that the reported Henry's law constants in the literature can vary over a wide range, depending on the research source. This range can be an order of magnitude or more, so different sources should be consulted. The same is true for the temperature factors reported in the literature. For this reason, the efficiency of any gas transfer equipment should be determined in real application or piloting over the temperature range of the application. Manufacturers often test their equipment designs to generate design data so that they can accurately predict performance for different applications.

Dimensionless Henry's constants are frequently used in engineering design. However, Henry's law constants reported in the literature can be reported in a number of units including but not limited to atm, atm-L/mole, bar, and bar-L/mole. Conversion from units reported in the literature is often required; Table 2-2 provides a number of calculations that can be used to convert from commonly reported units to dimensionless units. Table 2-3 provides dimensionless Henry's law constants for a number of organic compounds and gases that can be encountered in drinking water treatment.

In the event that Henry's Law constants are not readily available, the Henry's law constant can be estimated. However, the estimation technique only provides values within ± 50 to 100 percent of experimental values and should therefore only be used if measured values are not available. The following equation can be used to estimate the Henry's law constant when the mole fraction of the compound is equal to 1 and the compound is perfectly miscible in solution (Hand 1999):

$$K_H = P_{v,A} \tag{Eq 2-2}$$

where $P_{V,A}$ is the vapor pressure of compound A at a specified temperature (atm).

Variations in ionic strength and temperature may change the Henry's law constant. As temperature increases, the solubility of gases in water decreases and the partial pressure of a specific compound in the gas phase increases, thereby increasing the Henry's law constant. As temperature decreases, the solubility of gases in water increases and the partial pressure of a specific compound in the gas phase decreases, thereby decreasing the Henry's law constant. Temperature effects are generally driven by the water temperature rather than the air temperature, as air rapidly approaches the water temperature during initial contact, due to the relatively high heat capacity of water compared to air.

Solution pH affects the distribution of species between un-ionized and ionized forms, which influences the volatility of compounds. Only un-ionized species (e.g., NH_3 rather than NH_4^+ and CO_2 rather than HCO_3^-) are volatile, which makes them amenable to air–water equilibrium partitioning.

Table 2-2 Henry's law constant conversions

Units for Henry's Law Constant	Conversion to Dimensionless Henry's Law Constant	Forms of Universal Gas Constant
atm	$= \dfrac{K_H[atm]}{RT \times 55.56\dfrac{mol}{L}}$	$R = 0.08205\dfrac{atm \cdot L}{mol \cdot K}$
$\dfrac{atm \cdot L}{mol}$	$= \dfrac{K_H\left[\dfrac{atm \cdot L}{mol}\right]}{RT}$	$R = 0.08205\dfrac{atm \cdot L}{mol \cdot K}$
bar	$= \dfrac{K_H[bar]}{RT \times 55.56\dfrac{mol}{L}}$	$R = 0.083145\dfrac{bar \cdot L}{mol \cdot K}$
$\dfrac{bar \cdot L}{mol}$	$= \dfrac{K_H\left[\dfrac{bar \cdot L}{mol}\right]}{RT}$	$R = 0.083145\dfrac{bar \cdot L}{mol \cdot K}$

T = temperature [K]

Table 2-3 Henry's law constants for select volatile organic compounds and gases encountered in drinking water treatment

Component	KH (-)	T (°C)	Source
1,1,1-trichloroethane	0.416	10	Hand 1999, Ashworth 1988
	0.487	15	Hand 1999, Ashworth 1988
	0.607	20	Hand 1999, Ashworth 1988
	0.712	25	Hand 1999, Ashworth 1988
	0.849	30	Hand 1999, Ashworth 1988
1,1,2,2-tetrachloroethane	0.014	10	Hand 1999, Ashworth 1988
	0.008	15	Hand 1999, Ashworth 1988
	0.030	20	Hand 1999, Ashworth 1988
	0.010	25	Hand 1999, Ashworth 1988
	0.028	30	Hand 1999, Ashworth 1988
1,1,2-trichloroethane	0.017	10	Hand 1999, Ashworth 1988
	0.027	15	Hand 1999, Ashworth 1988
	0.031	20	Hand 1999, Ashworth 1988
	0.037	25	Hand 1999, Ashworth 1988
	0.054	30	Hand 1999, Ashworth 1988
1,1,2-trichlorotrifluoroethane	6.63	10	Hand 1999, Ashworth 1988
	9.1	15	Hand 1999, Ashworth 1988
	10.2	20	Hand 1999, Ashworth 1988
	13	25	Hand 1999, Ashworth 1988
	12.9	30	Hand 1999, Ashworth 1988
1,1-dichloroethane	0.158	10	Hand 1999, Ashworth 1988
	0.192	15	Hand 1999, Ashworth 1988
	0.234	20	Hand 1999, Ashworth 1988
	0.256	25	Hand 1999, Ashworth 1988
	0.312	30	Hand 1999, Ashworth 1988
1,1-dichloroethylene	0.663	10	Hand 1999, Ashworth 1988
	8.59	15	Hand 1999, Ashworth 1988
	0.907	20	Hand 1999, Ashworth 1988
	1.06	25	Hand 1999, Ashworth 1988
	1.28	30	Hand 1999, Ashworth 1988
1,2,4-trichiorobenzene	0.056	10	Hand 1999, Ashworth 1988
	0.044	15	Hand 1999, Ashworth 1988
	0.076	20	Hand 1999, Ashworth 1988
	0.079	25	Hand 1999, Ashworth 1988
	0.119	30	Hand 1999, Ashworth 1988
1,2-dichloroethane	0.050	10	Hand 1999, Ashworth 1988
	0.055	15	Hand 1999, Ashworth 1988
	0.061	20	Hand 1999, Ashworth 1988
	0.058	25	Hand 1999, Ashworth 1988
	0.07	30	Hand 1999, Ashworth 1988

Continued

Table 2-3 Henry's law constants for select volatile organic compounds and gases encountered in drinking water treatment *(continued)*

Component	KH (-)	T (°C)	Source
1,2-dichloropropane	0.053	10	Hand 1999, Ashworth 1988
	0.053	15	Hand 1999, Ashworth 1988
	0.079	20	Hand 1999, Ashworth 1988
	0.146	25	Hand 1999, Ashworth 1988
	0.115	30	Hand 1999, Ashworth 1988
1,3,5-trimethylbenzene	0.174	10	Hand 1999, Ashworth 1988
	0.195	15	Hand 1999, Ashworth 1988
	0.238	20	Hand 1999, Ashworth 1988
	0.275	25	Hand 1999, Ashworth 1988
	0.387	30	Hand 1999, Ashworth 1988
1,4-dichlorobenzene	0.091	10	Hand 1999, Ashworth 1988
	0.092	15	Hand 1999, Ashworth 1988
	0.108	20	Hand 1999, Ashworth 1988
	0.13	25	Hand 1999, Ashworth 1988
	0.156	30	Hand 1999, Ashworth 1988
2,4-dimethylphenol	0.357	10	Hand 1999, Ashworth 1988
	0.285	15	Hand 1999, Ashworth 1988
	0.42	20	Hand 1999, Ashworth 1988
	0.202	25	Hand 1999, Ashworth 1988
	0.151	30	Hand 1999, Ashworth 1988
2-methylpentane	30	10	Hand 1999, Ashworth 1988
	29.4	15	Hand 1999, Ashworth 1988
	26.3	20	Hand 1999, Ashworth 1988
	33.7	25	Hand 1999, Ashworth 1988
	34.1	30	Hand 1999, Ashworth 1988
1,2-dichlorobenzene	0.070	10	Hand 1999, Ashworth 1988
	0.061	15	Hand 1999, Ashworth 1988
	0.070	20	Hand 1999, Ashworth 1988
	0.064	25	Hand 1999, Ashworth 1988
	0.095	30	Hand 1999, Ashworth 1988
1,3-dichlorobenzene	0.095	10	Hand 1999, Ashworth 1988
	0.098	15	Hand 1999, Ashworth 1988
	0.122	20	Hand 1999, Ashworth 1988
	0.117	25	Hand 1999, Ashworth 1988
	0.17	30	Hand 1999, Ashworth 1988
ammonia	0.001	20	Hand 1999, Montgomery 1985
benzene	0.142	10	Hand 1999, Ashworth 1988
	0.164	15	Hand 1999, Ashworth 1988
	0.188	20	Hand 1999, Ashworth 1988
	0.216	25	Hand 1999, Ashworth 1988
	0.29	30	Hand 1999, Ashworth 1988

Continued

Table 2-3 Henry's law constants for select volatile organic compounds and gases encountered in drinking water treatment *(continued)*

Component	KH (-)	T (°C)	Source
carbon dioxide	1.14	20	Hand 1999, Montgomery 1985
carbon tetrachloride	0.637	10	Hand 1999, Ashworth 1988
	0.808	15	Hand 1999, Ashworth 1988
	0.965	20	Hand 1999, Ashworth 1988
	1.21	25	Hand 1999, Ashworth 1988
	1.52	30	Hand 1999, Ashworth 1988
chlorine	0.442	20	Hand 1999, Montgomery 1985
chlorine dioxide	0.041	20	Hand 1999, Montgomery 1985
chloroethane	0.327	10	Hand 1999, Ashworth 1988
	0.405	15	Hand 1999, Ashworth 1988
	0.458	20	Hand 1999, Ashworth 1988
	0.495	25	Hand 1999, Ashworth 1988
	0.575	30	Hand 1999, Ashworth 1988
chlorobenzene	0.105	10	Hand 1999, Ashworth 1988
	0.119	15	Hand 1999, Ashworth 1988
	0.142	20	Hand 1999, Ashworth 1988
	0.147	25	Hand 1999, Ashworth 1988
	0.19	30	Hand 1999, Ashworth 1988
chloroform	0.074	10	Hand 1999, Ashworth 1988
	0.986	15	Hand 1999, Ashworth 1988
	0.138	20	Hand 1999, Ashworth 1988
	0.172	25	Hand 1999, Ashworth 1988
	0.223	30	Hand 1999, Ashworth 1988
cis- 1,2-dichloroethylene	0.116	10	Hand 1999, Ashworth 1988
	0.138	15	Hand 1999, Ashworth 1988
	0.15	20	Hand 1999, Ashworth 1988
	0.186	25	Hand 1999, Ashworth 1988
	0.231	30	Hand 1999, Ashworth 1988
cyclohexane	4.44	10	Hand 1999, Ashworth 1988
	5.33	15	Hand 1999, Ashworth 1988
	5.82	20	Hand 1999, Ashworth 1988
	7.24	25	Hand 1999, Ashworth 1988
	8.97	30	Hand 1999, Ashworth 1988
dibromochloromethane	0.016	10	Hand 1999, Ashworth 1988
	0.019	15	Hand 1999, Ashworth 1988
	0.043	20	Hand 1999, Ashworth 1988
	0.048	25	Hand 1999, Ashworth 1988
	0.061	30	Hand 1999, Ashworth 1988

Continued

Table 2-3 Henry's law constants for select volatile organic compounds and gases encountered in drinking water treatment *(continued)*

Component	KH (-)	T (°C)	Source
ethylbenzene	0.14	10	Hand 1999, Ashworth 1988
	0.191	15	Hand 1999, Ashworth 1988
	0.25	20	Hand 1999, Ashworth 1988
	0.322	25	Hand 1999, Ashworth 1988
	0.422	30	Hand 1999, Ashworth 1988
ethylene dibromide	0.013	10	Hand 1999, Ashworth 1988
	0.020	15	Hand 1999, Ashworth 1988
	0.025	20	Hand 1999, Ashworth 1988
	0.027	25	Hand 1999, Ashworth 1988
	0.032	30	Hand 1999, Ashworth 1988
hydrogen sulfide	0.389	20	Hand 1999, Montgomery 1985
m-xylene	0.177	10	Hand 1999, Ashworth 1988
	0.21	15	Hand 1999, Ashworth 1988
	0.249	20	Hand 1999, Ashworth 1988
	0.304	25	Hand 1999, Ashworth 1988
	0.357	30	Hand 1999, Ashworth 1988
methane	28.7	20	Hand 1999, Montgomery 1985
methyl cellosolve	1.9	10	Hand 1999, Ashworth 1988
	1.54	15	Hand 1999, Ashworth 1988
	4.83	20	Hand 1999, Ashworth 1988
	1.26	25	Hand 1999, Ashworth 1988
	1.53	30	Hand 1999, Ashworth 1988
methylene chloride	0.060	10	Hand 1999, Ashworth 1988
	0.072	15	Hand 1999, Ashworth 1988
	0.102	20	Hand 1999, Ashworth 1988
	0.121	25	Hand 1999, Ashworth 1988
	0.145	30	Hand 1999, Ashworth 1988
methyl ethylbenzene	0.151	10	Hand 1999, Ashworth 1988
	0.178	15	Hand 1999, Ashworth 1988
	0.209	20	Hand 1999, Ashworth 1988
	0.228	25	Hand 1999, Ashworth 1988
	0.31	30	Hand 1999, Ashworth 1988
methyl ethyl ketone	0.012	10	Hand 1999, Ashworth 1988
	0.017	15	Hand 1999, Ashworth 1988
	0.008	20	Hand 1999, Ashworth 1988
	0.005	25	Hand 1999, Ashworth 1988
	0.004	30	Hand 1999, Ashworth 1988

Continued

Table 2-3 Henry's law constants for select volatile organic compounds and gases encountered in drinking water treatment *(continued)*

Component	KH (-)	T (°C)	Source
methyl isobutyl ketone	0.028	10	Hand 1999, Ashworth 1988
	0.016	15	Hand 1999, Ashworth 1988
	0.012	20	Hand 1999, Ashworth 1988
	0.016	25	Hand 1999, Ashworth 1988
	0.027	30	Hand 1999, Ashworth 1988
n-hexane	10.3	10	Hand 1999, Ashworth 1988
	17.5	15	Hand 1999, Ashworth 1988
	36.7	20	Hand 1999, Ashworth 1988
	31.4	25	Hand 1999, Ashworth 1988
	62.7	30	Hand 1999, Ashworth 1988
nonane	17.2	10	Hand 1999, Ashworth 1988
	21	15	Hand 1999, Ashworth 1988
	13.8	20	Hand 1999, Ashworth 1988
	16.9	25	Hand 1999, Ashworth 1988
	18.7	30	Hand 1999, Ashworth 1988
o-xylene	0.123	10	Hand 1999, Ashworth 1988
	0.153	15	Hand 1999, Ashworth 1988
	0.197	20	Hand 1999, Ashworth 1988
	0.199	25	Hand 1999, Ashworth 1988
	0.252	30	Hand 1999, Ashworth 1988
oxygen	32.5	20	Hand 1999, Montgomery 1985
ozone	3.77	20	Hand 1999, Montgomery 1985
p-xylene	0.181	10	Hand 1999, Montgomery 1985
	0.204	15	Hand 1999, Montgomery 1985
	0.268	20	Hand 1999, Montgomery 1985
	0.304	25	Hand 1999, Montgomery 1985
	0.38	30	Hand 1999, Montgomery 1985
propylbenzene	0.245	10	Hand 1999, Ashworth 1988
	0.309	15	Hand 1999, Ashworth 1988
	0.366	20	Hand 1999, Ashworth 1988
	0.442	25	Hand 1999, Ashworth 1988
	0.551	30	Hand 1999, Ashworth 1988
radon	1.69	20	Hand 1999, Montgomery 1985
sulfur dioxide	0.029	20	Hand 1999, Montgomery 1985
tetrachloroethylene	0.364	10	Hand 1999, Ashworth 1988
	0.467	15	Hand 1999, Ashworth 1988
	0.587	20	Hand 1999, Ashworth 1988
	0.699	25	Hand 1999, Ashworth 1988
	0.985	30	Hand 1999, Ashworth 1988

Continued

Table 2-3 Henry's law constants for select volatile organic compounds and gases encountered in drinking water treatment *(continued)*

Component	KH (-)	T (°C)	Source
toluene	0.164	10	Hand 1999, Ashworth 1988
	0.21	15	Hand 1999, Ashworth 1988
	0.231	20	Hand 1999, Ashworth 1988
	0.263	25	Hand 1999, Ashworth 1988
	0.325	30	Hand 1999, Ashworth 1988
trans- 1,2-dichloroethylene	0.025	10	Hand 1999, Ashworth 1988
	0.298	15	Hand 1999, Ashworth 1988
	0.356	20	Hand 1999, Ashworth 1988
	0.386	25	Hand 1999, Ashworth 1988
	0.469	30	Hand 1999, Ashworth 1988
trichloroethylene	0.237	10	Hand 1999, Ashworth 1988
	0.282	15	Hand 1999, Ashworth 1988
	0.350	20	Hand 1999, Ashworth 1988
	0.417	25	Hand 1999, Ashworth 1988
	0.515	30	Hand 1999, Ashworth 1988
trichlorofluoromethane	2.31	10	Hand 1999, Ashworth 1988
	2.88	15	Hand 1999, Ashworth 1988
	3.34	20	Hand 1999, Ashworth 1988
	4.13	25	Hand 1999, Ashworth 1988
	4.91	30	Hand 1999, Ashworth 1988
vinyl chloride	0.646	10	Hand 1999, Ashworth 1988
	0.711	15	Hand 1999, Ashworth 1988
	0.903	20	Hand 1999, Ashworth 1988
	1.08	25	Hand 1999, Ashworth 1988

Note: Refer to "Other References" in Chapter 1 if compound of interest is not found in this table.

Surfactants, if present, would decrease the Henry's law constant of the compounds of interest (thereby decreasing ability to volatilize). In drinking water applications, water supplies do not typically have concentrations of surfactants that would affect the design of air stripping equipment.

If literature values are not available for a specific temperature, the van 't Hoff equation can be used to calculate the change in K_H with temperature. For the temperatures encountered in water treatment, K_H generally increases with increasing temperature because the aqueous solubility of the component decreases while its vapor pressure increases.

$$\log K_H = \frac{-\Delta H^0}{RT} + E \quad \text{(Eq 2-3)}$$

where:

K_H = calculated Henry's law constant

$-\Delta H^0$ = heat absorbed in the evaporation of 1 mole of component A from solution at constant temperature and pressure

R = universal gas constant $\left(1.987 \frac{kcal}{kmol \cdot K}\right)$

T = temperature

E = empirical constant

The standard enthalpy change, ΔH, for the dissolution of a compound in water is assumed to be constant over the temperature range of interest in the previous equation.

Tables 2-4 and 2-5 include values of ΔH and E for several volatile organic compounds and gases.

Temperature dependence of the Henry's law constant can also be expressed using the following temperature regression equation:

$$K_H = \exp\left(A - \frac{B}{T}\right) \qquad \text{(Eq 2-4)}$$

where A and B are fitting parameters, T is temperature (K), and K_H is the Henry's law constant (atm-m^3/gmole).

Table 2-6 provides values of the fitting parameters for a selection of compounds that can be encountered in drinking water treatment. Note that this table is not comprehensive, so other references, such as those listed in Chapter 1, may be needed if a compound is not listed in this table.

Table 2-4 Henry's constants and temperature dependence for select volatile organic compounds in water

Gas	Molecular Formula	Henry's Constant at 20°C (atm)*	Temperature Dependence	
			ΔH^0 (kcal/kmol, × 10^{-3})	E
Carbon tetrachloride	CCl_4	1.29×10^3	4.05	10.06
Tetrachloroethylene	C_2Cl_4	1.10×10^3	4.29	10.38
Trichloroethylene	$CCHCl_3$	550	3.41	8.59
Benzene	C_6H_6	240	3.68	8.68
Chloroform	$CHCl_3$	170	4.00	9.10

* Calculated from water solubility data and pure liquid partial pressure at specified temperature (Landolt-Boernstein 1976, Montgomery 1985)

Note: Refer to "Resources" in Chapter 1 if compound of interest is not found in this table.

Table 2-5 Henry's constants and temperature dependence for gases in water

Gas	Molecular Formula	Henry's Constant at 20°C (atm)	Temperature Dependence	
			ΔH^0 (kcal/kmol, × 10^{-3})	E
Oxygen	O_2	4.3×10^4	1.45	7.11
Nitrogen	N_2	8.6×10^4	1.12	6.85
Methane	CH_4	3.8×10^4	1.54	7.22
Ozone	O_3	5.0×10^3	2.52	8.05
Carbon dioxide	CO_2	1.51×10^2	2.07	6.73
Hydrogen sulfide	H_2S	5.15×10^2	1.85	5.88
Chlorine	Cl_2	5.85×10^2	1.74	5.75
Chlorine dioxide	ClO_2	54	2.93	6.76
Sulfur dioxide	SO_2	38	2.40	5.68
Ammonia	NH_3	0.76	3.75	6.31

(Landolt-Boernstein 1976 and International Critical Tables, Montgomery 1985)

Note: Refer to "Resources" in Chapter 1 if compound of interest is not found in this table.

Table 2-6 Fitting parameters for the temperature regression equation

Component	A	B	R^2
1,2-dichloropropane	9.843	4,708	0.820
1,3-dichlorobenzene	2.882	2,564	0.850
1,1,2-trichlorotrifluoroethane	9.649	3,243	0.932
1,1,1-trichloroethane	7.351	3,399	0.998
1,1,2-trichloroethane	9.320	4,843	0.968
1,1-dichloroethane	5.484	3,137	0.993
1,1-dichloroethylene	6.123	2,907	0.974
1,2,4-trichlorobenzene	7.361	4,028	0.819
1,2-dichloroethane	−1.371	1,522	0.878
1,3,5-trimethylbenzene	7.241	3,628	0.962
1,4-dichlorobenzene	3.373	2,720	0.941
Benzene	5.534	3,194	0.968
carbon tetrachloride	9.739	3,951	0.997
Chlorobenzene	3.469	2,689	0.965
Chloroethane	4.265	2,580	0.984
Chloroform	11.41	5,030	0.997
cis-1,2-dichloroethylene	5.164	3,143	0.974
Cyclohexane	9.141	3,238	0.982
Decalin	11.85	4,125	0.919
dibromochloromethane	14.62	6,373	0.914
Ethylbenzene	11.92	4,994	0.999
ethylene dibromide	5.703	3,876	0.928
m-xylene	6.280	3,337	0.998
methylene chloride	8.483	4,268	0.988
methyl ethylbenzene	5.557	3,179	0.968
methyl ethyl ketone	−26.32	−5,214	0.797
n-hexane	25.25	7,530	0.917
o-xylene	5.541	3,220	0.966
p-xylene	6.931	3,520	0.989
Propylbenzene	7.835	3,681	0.997
tetrachloroethylene	10.65	4,368	0.987
Tetralin	11.83	5,392	0.996
Toluene	5.133	3,024	0.982
trans-1,2-dichloroethylene	5.333	2,964	0.985
trichloroethylene	7.845	3,702	0.998
trichlorofluoromethane	9.480	3,513	0.998
vinyl chloride	6.138	2,931	0.970

Source: Ashworth 1988
Note: Refer to "Resources" in Chapter 1 if compound of interest is not found in this table.

GAS TRANSFER

Film Diffusion

Transfer of material in air stripping and aeration applications occurs from one bulk phase to another across a gas–liquid interface. The portion of each media where concentrations differ from the bulk phase is referred to as the boundary layer, and transfer from one bulk media to another is often conceptualized as having to overcome a resistance in a thin film in the boundary layer of each phase. These liquid and gas film resistances control the rate of mass transfer. In general, the liquid film has a greater influence on the transfer of low solubility gases like carbon dioxide and oxygen. Turbulence and temperature are the most important factors in film resistance. Gas transfer is enhanced as temperature and turbulence increase (thereby reducing the boundary layer thickness and increasing rate of molecular diffusion).

The differential between the saturation value of a compound (that is its equilibrium concentration as described by Henry's law) and its actual concentration in the water or air provides the driving force for phase transfer. The phase transfer occurs in the direction of decreasing the difference between the actual concentration and the saturation concentration. In air stripping applications, the concentration of the target compound in water must exceed the saturation concentration. The more that the concentration exceeds the saturation concentration, the greater the driver for phase transfer from the water phase to air phase. In aeration applications, the concentration of the target compound in water must be below saturation, and the concentration of a gas in air, the greater driver for phase transfer from the air phase to water phase.

Gas Transfer Without Chemical Reaction

The overall resistance to mass transfer can be described as the sum of the liquid-phase resistance and gas-phase resistance to mass transfer. The overall resistance to mass transfer is described as follows:

$$\frac{1}{K_L a} = \frac{1}{k_l a} + \frac{1}{K_H k_g a} \qquad \text{(Eq 2-5)}$$

where:

K_L = overall liquid phase mass transfer coefficient (ft/h) [m/h]
k_l = liquid phase mass transfer coefficient (ft/h) [m/h]
k_g = gas phase mass transfer coefficient (ft/h) [m/h]
K_H = Henry's law constant
a = effective area for mass transfer (ft^2) [m^2]

The effective area for mass transfer, a, describes the interfacial area of the system in which mass transfer occurs. The mass transfer coefficient, K_L, is a function of the compound's Henry's law constant, the geometry and physical characteristics of the air stripping system, system temperature, contact time, and liquid-phase and gas-phase resistances to mass transfer (k_g and k_l).

For gas scrubbing operations, the previous equation can also be written to describe the overall mass transfer coefficient for the gas phase:

$$\frac{1}{K_G a} = \frac{K_H}{k_l a}\frac{K_H}{k_l a} + \frac{1}{k_g a} \qquad \text{(Eq 2-6)}$$

where:

K_G = overall gas phase mass transfer coefficient (L/T)
a = effective area for mass transfer (ft^2) [m^2]

K_H = Henry's law constant
k_l = liquid phase mass transfer coefficient (ft/h) [m/h]
k_g = gas phase mass transfer coefficient (ft/h) [m/h]

The product of the mass transfer coefficient, K_L or K_G, and the effective area for mass transfer, a, is commonly evaluated as one constant, $K_L\,a$ or $K_G\,a$.

In turbulent flow regimes, mass transfer coefficients must be determined experimentally. When experimental data are not available, mass transfer correlations can be used for the design of various aeration and air stripping equipment.

The volumetric rate of transfer of a volatile compound from water to air can be expressed by the following first-order equation:

$$M = K_L a\left(C_i^* - D_i\right) \qquad \text{(Eq 2-7)}$$

where:

M = mass of substance transferred per unit of time per unit of volume (lb/h)/ft^3 [kg/h)/m^3]
K_L = overall liquid mass transfer coefficient (ft/h) [m/h]
a = effective area for mass transfer (ft^2/ft^3) [m^2/m^3]
C_i^* = liquid-phase concentration in equilibrium with gas-phase concentration (lb/f^{t3}) [kg/m^3]
D_i = bulk liquid-phase concentration (lb/ft^3) [kg/m^3]

The rate of mass transfer is a function of the difference between the existing concentration and the equilibrium concentration of the substance in solution. Increasing the effective area for mass transfer will increase the mass transfer rate; however, capital and operating costs will also increase as more surface area will be required (e.g., additional packing). Designers typically design aeration equipment to optimize mass transfer while balancing capital and energy costs.

The mass transfer rate will be controlled by either the liquid or gas phase. Understanding the controlling phase is important for design and operations refinement. A concentration gradient is required to exist in the direction of transfer of each phase, so designers and operators need to know the controlling phase in order to keep the concentration gradient moving in the controlling direction. Table 2-7 identifies the controlling phase for select VOCs and gases. The following rule of thumb can be used to determine which phase controls the mass transfer rate for a compound:

- $K_H > 0.05$ and low liquid solubility: liquid-phase controls
- $K_H < 0.002$ and high liquid solubility: gas-phase controls
- $0.0002 < K_H < 0.05$: both liquid and gas phase may affect mass transfer rate

Table 2-7 Controlling phase for select compounds

Compound	K_H (25°C)	Controlling Phase
Oxygen	29.29	Liquid
Carbon tetrachloride	1.2	Liquid
Trichloroethylene	0.40	Liquid
Chloroform	0.15	Liquid
Acetone	0.001	Gas

Source: Hand 2011

Gas Transfer With Chemical Reaction

Chemical reactions can occur in gas transfers through the gas–water interface resulting in an increased absorption rate. Gases used in the water treatment industry that can undergo chemical reactions include sulfur dioxide, carbon dioxide, ozone, and chlorine. Design of aeration equipment needs to account for these chemical reactions.

The impact on the rate of absorption due to chemical reactions is a function of the chemical reaction type, the reactant concentrations, rate constant, reaction order, and solute molecular diffusion constants. Montgomery (1985), Danckwerts (1970), and Sherwood (1975) present methods for calculating the volumetric rate of mass transfer.

It is important to emphasize that there are removal and addition limitations when air is used to bring about gas transfer. The atmosphere is made up of several natural gases, and in the vicinity of contamination sites, air can contain various amounts of other volatile contaminants. To the extent that these atmospheric gases are present, removal and addition are limited by the equilibration given by Henry's constant. In other words, if the air used in a process contains that which is being transferred, it is not possible to remove all of the constituents. Of interest with regard to drinking water is that air always naturally contains oxygen, carbon dioxide, and radon. At a given water temperature, the saturation of these gases with respect to the air–water equilibrium defines the concentrations that will remain with any amount of aeration or air stripping. For radon, this will likely be a few tenths to 0.5 pCi/L (average air concentration near ground of 1.5 pCi/L); for oxygen, it is in the range of ≈ 8–14 mg/L; and for carbon dioxide, it is in the range of 0.4–0.8 mg/L. These values are meant to cover the range of typical drinking water temperatures.

Air containing the constituent that is being transferred is most important in the location of the blower air filter intake relative to the discharging air vent. It is not unheard of that a partial return of vent air comes back into a blower and limits the removal efficiency. A reasonable separation of the two air streams is needed to ensure that this does not occur and typically 10 feet, with opposing direction for the air flows, is sufficient to avoid this potential problem.

GAS TRANSFER EQUIPMENT

This section discusses the various types of gas addition and removal technologies and their general application, design, and operational considerations. The order of presentation of equipment types is based on the general historical chronology of development.

Some of the first techniques used to aerate water employed a simple cascade or waterfall. It was recognized early on that water was "freshened" by cascading it over rocks, wooden slats, and sequential trays. This practice was an attempt to equilibrate the water with air, to add oxygen, to remove carbon dioxide and hydrogen sulfide, and to start the process of iron oxidation as a prelude to filtration of the precipitated iron. As the process was developed, coke trays were often used to improve the removal of manganese, which much later was found to need a manganese dioxide coating on surfaces to carry out oxidation and removal on filters. Coke tray aerators are a series of activated carbon trays where water passes through each plate through small holes. This type of aeration also raised the pH and reduced corrosivity for waters that had significant carbon dioxide. The majority of these operations focused on groundwater treatment. The "Resources" section in Chapter 1 lists reference materials that offer more information on the historical aspects of aeration.

The development of more sophisticated equipment that was designed for specific reactor applications requiring much higher removal efficiency largely occurred in the 1940s through the 1960s in the field of chemical engineering. The equipment included packed towers, spray devices, and multi-staged bubble reactors in both horizontal and in stacked tray configurations. During the 1970s, while USEPA and states began establishing maximum concentration levels (MCLs) for a variety of VOCs, engineers in the drinking water field began using the more efficient equipment. Packed towers were the most common equipment applied to the removal of VOCs, both in the groundwater remediation effort associated with the Superfund program and leaking underground fuel storage and in drinking water supplies that were contaminated. At the same time, lower-profile air strippers in the form of multi-staged bubble equipment were used as alternatives to packed towers. These low-profile systems were easier to locate inside buildings and did not cause an industrial-looking installation in more highly populated areas. In addition, there were some advantages for easier maintenance. The staged bubble systems used two basic designs: (1) a cascade of vertical trays employing countercurrent air:water flow, and (2) a staged horizontal vessel using air diffusers in each stage with blower air distributed evenly to all stages. Packed towers and low-profile air strippers can be designed to achieve almost complete removal of volatile contaminants with a single pass.

It should be noted that other types of less efficient equipment, including venturis/eductors, spray, cascades, etc., can also attain high removals, but they would often require multiple-pass configurations that are not cost-effective when isolated batch processing and re-pumping requirements are included.

There are specific applications in which the lower removal devices offer practical advantages and are employed in drinking water treatment. One example is where only a low degree of removal is required to meet the treatment goal (see Table 2-8). A second example is where an existing deep storage tank is used for the removal of THHMs (see Chapter 4). In this case, the energy required for a blower or compressor is cost prohibitive because of excess depth, and a surface or a spray device with continuous circulation may be sufficient to provide a cost-effective treatment. This may be more economical in some cases, such as when the cost of an additional air stripper after storage and a possible re-pumping step is required. Lastly, for very small water systems, an educator or a venturi device is sometimes used for the addition of a small amount of oxygen to oxidize dissolved iron.

Finally, many of the installations of gas transfer equipment used in drinking water are predesigned by the manufacturer. The design engineer works with the manufacturer to select the appropriate capacity system and ensure that the unit selected will perform as specified.

Weir, Cascade, and Tray Aerators

In weir, cascade, and tray aerators, water drops by gravity onto several steps. Each step contains an overflow weir and a receiving gutter. When water passes over a weir, an interface between air and water is created. When the falling water submerges into the receiving body of water, significant amounts of air are entrained. The entrained air is then dispersed in the form of bubbles throughout the receiving body of water, which leads to the transfer of gases. Figure 2-1 shows a schematic of a typical cascade aerator.

Table 2-8 Efficiency coefficient k of different gases as a function of the weir height, in percent

Gas	h = 1 ft (0.4m)	h = 2 ft (0.8 m)	h = 4 ft (1.2)
O_2	25	46	55
CO_2	14	15	15
CH_4	27	48	62

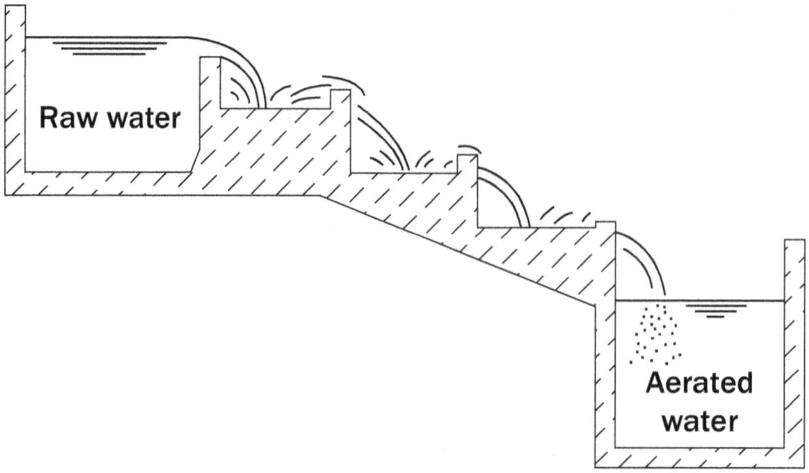

Source: Appropedia article entitled "Original: Slow sand filtration water treatment plants"

Figure 2-1 Schematic of a cascade aerator

The gas transfer takes place at the interface between the water and the air bubbles. Because the amount of air that is entrained is limited, the air-to-water ratio is also limited, typically about 0.4 to 1.

An estimate of the efficiency for a cascade can be made, assuming that there is a relationship between the measured fall height and the efficiency. The efficiency of a cascade depends on the fall height of each cascade step and the number of steps:

$$K = \frac{c_{we} - c_{wo}}{C_S - c_{wo}} = 1 - (1 - k)^n \quad \text{(Eq 2-8)}$$

where:

K = overall efficiency of the cascade
c_{w0} = initial concentration of gas in water [g/m^3]
c_{we} = final concentration of gas in water [g/m^3]
C_s = concentration of gas in water at saturation [g/m^3]
k = efficiency for each step [%]
n = number of steps

In Table 2-8, the efficiency is given for oxygen, carbon dioxide, and methane as a function of the fall height of a step. With the data from Table 2-8 and Equation 2-8, the efficiency of a cascade with n number of steps can be calculated. In practice, the total fall height of all the cascade steps together typically varies between 5 to 20 feet (2 to 7 m).

Packed Towers

Packed tower aeration (PTA) involves passing water down through a column of packing material while forcing air up through the packing media as shown in Figure 2-2.

Generally, aeration is feasible for removing compounds with a Henry's law constant greater than 100 (expressed in atm mol/mol) but not normally feasible for removing compounds with a Henry's constant of less than 10. For values between 10 and 100, PTA or other forced-air systems are used but should be evaluated using pilot studies or modeling. Table 2-9 shows categories of compounds in water classified by ease of aeration.

Because the Henry's law constant for each compound decreases with decreasing temperature, the system should be designed for the coldest air and water temperatures expected during operation.

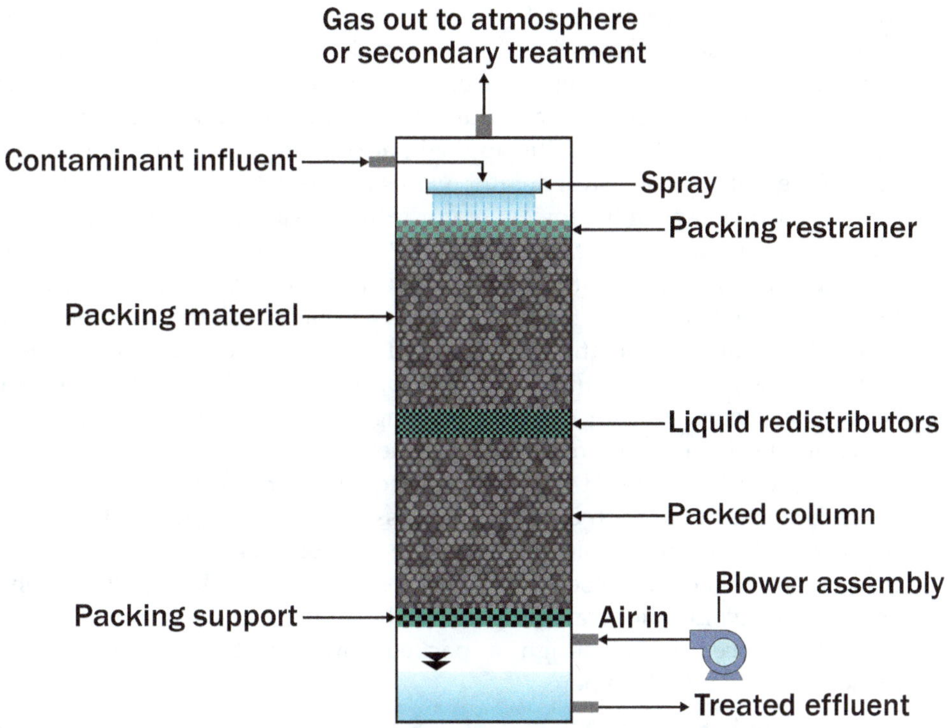

Source: USEPA

Figure 2-2 Schematic of a packed tower aerator

Table 2-9 Aerated compounds in water (based on Henry's constant at 20°C)

Henry's Constant	Compounds
>100; readily aerated	Vinyl chloride, nitrogen, methane, ozone, toxaphene, carbon dioxide, radon, carbon tetrachloride, tetrachloroethylene, trichloroethylene, hydrogen sulfide, chloromethane, 1,1,1-trichloroethylene, toluene, 1,2,4-trimethylbenzene, benzene, 1,4-dichlorobenzene, chloroform
>10 and <100; potentially removed with packed tower aeration	1-2-dichloromethane, 1,1,2-trichloroethene, sulfide dioxide, bromoform
<10; not readily aerated	Ammonia, oxygen, pentachlorphenol, dieldrin, benzene, aldicarb, chlordane, polychlorinated biphenols

Source: Odell 2010

A tower aerator consists of a cylinder of steel or synthetic material filled with a packing medium (see Figure 2-2). Packing media can consist of stacked slats or tubes, or specially designed packing material like the Pall-ring and the Berl-saddle. In the top section of the tower, the water is divided over the packing media and flows down over the media surface. As a result of the flow of water over the packing media, a large contact surface between the air and water is created for gas transfer. In addition, the water falls in drops from one packing element to the other, continuously forming new drops and renewing the air-water interface for additional gas transfer (Delft University of Technology 2019).

The air can be renewed by natural ventilation or with the help of a ventilator. If a ventilator is used, the air can have a co- or counter-current flow in the tower, as a ventilator can help to distribute the air if it is in a co-current flow with water. With natural ventilation, the air is required to be counter-current with water. The packing material can be produced from synthetic material, metal, carbon, or ceramic material. The dimensions of the individual pieces vary from 6 mm to 75 mm. In practice, installations used for purifying drinking water use mostly synthetic packing material with a dimension of 25–50 mm.

The surface loading (flow divided by surface area) that in practice is used in tower aerators is 40–100 $m^3/(m^2{\cdot}h)$. The applied packing height, which determines the retention time of the water in the tower aerator, varies between 3 and 5 m.

A disadvantage of the tower aerator is that the system is sensitive to clogging. If iron (Fe^{2+}) is present in groundwater, it will oxidize in the tower aerator (Fe^{3+}) and accumulate on the packing material as iron hydroxide ($Fe(OH)_3$). Because the oxidized iron influences the gas transfer negatively, it will be necessary to back flush the tower aerator. Water with a high velocity, or a combination of water and air, is then flushed through the tower aerator, removing the iron contamination from the packing material. In addition to flushing, it will be necessary to periodically clean the packing material chemically. In this case, the packing material must be removed from the tower aerator.

CO_2 can be stripped and pH subsequently increased through air stripping. In groundwater that has high calcium hardness and alkalinity, this pH increase can result in conditions that result in calcium carbonate precipitation on the packing media. Periodic cleaning of the media or use of a sequestering chemical (like a polyphosphate) would be necessary to avoid solidification on the media and reduced gas transfer.

The procedure to design a packed tower system is described by El-Belhlil, El-Gezawiand Sbri (2012) here:

1. Identify a target contaminant or select the contaminant whose final water quality standard is the most difficult to achieve (usually the least volatile contaminant).
2. Select removal efficiency needed.
3. Select the lowest water temperature and determine Henry's law constant from the literature or calculations previously described in this chapter.
4. Select air-to-water (A/W) ratio: typical air-to-water ratios for groundwater are between 10 and 100. Compute the stripping factor *S*, where K_H' is the Henry's law constant for the selected contaminant at the lowest water temperature, Q_g is the gas flow rate, and Q_w is the water flow rate.

$$S = K_H' \frac{Q_g}{Q_w} \tag{Eq 2-9}$$

   Most designs used a stripping factor in the range 3 – 5. However, **S** as high as 10 has been used.

5. Compute the number of transfer units *NTU*, where $C_{L,in}$ is the concentration of the contaminant L into the packed tower and $C_{L,out}$ is the concentration of the contaminant L out of the packed tower.

$$NTU = \left(\frac{S}{S-1}\right)\ln\left[\frac{S}{S-1}\frac{C_{L,in}}{C_{L,out}} + \frac{1}{S}\right]\ln\left[\frac{S}{S-1}\frac{C_{L,in}}{C_{L,out}} + \frac{1}{S}\right] \tag{Eq 2-10}$$

6. Select a hydraulic loading rate and compute the cross-sectional area and the diameter of the tower. Hydraulic loading rates may vary between 5 and 50 gpm/ft^2 (3.4×10^{-3} to 0.034 $m^3/m^2{\cdot}s$). Hydraulic loading rates between 20 and 35 gpm/ft^2 are usually used.
7. Select the packing material and the mass transfer correlations to be used, such as Sherwood and Holloway's correlation (1940), which considers only the resistance in the liquid film, or Onda's correlation (1968), which includes both gas and liquid film resistances.

8. Compute the wetted surface area.

$$\frac{a_w}{a_t} = 1 - \exp\left[-1.45\left(\frac{\sigma_C}{\sigma_L}\right)^{0.75}\left(\frac{L_W^*}{a_t\mu_L}\right)^{0.1}\left(\frac{L_W^{*2}a_t}{\rho_L^2 g}\right)^{-0.05}\left(\frac{L_W^{*2}}{\rho_L\sigma_L a_t}\right)^{0.2}\right] \quad \text{(Eq 2-11)}$$

where:

a_w = effective interfacial area of packing per unit volume
a_t = actual area of packing per unit volume
σ_C = critical surface tension for the particular packing material
σ_L = liquid surface tension
L^*_w = liquid mass flow rate per unit cross-sectional area
ρ_L = density of water
μ_L = viscosity of water
g = gravity constant

The actual area of packing per unit volume (a_t) is usually obtained from the manufacturer, while the critical surface tension of the packing is assumed to be that of water. Several other pieces of information such as the density and viscosity of air and water can be obtained from standard textbooks or handbooks. Note that the equations use mass flux (kg/m^2·s).

9. If liquid and gas diffusivity of contaminant is not available, estimate the diffusivity of the contaminant in liquid and air phases with the Wilke and Chang (1955) equation and Hirschfelder, Bird, and Spotz (1946) equation, respectively.

10. With information from items 7 and 8, calculate the liquid and air phase mass transfer coefficients (k_L and k_G).

$$\frac{k_G}{a_t}\frac{RT}{D_G} = 5.23\left[\frac{V_W^*}{a_t\mu_G}\right]^{0.7}\left[\frac{\mu_G}{\rho_G D_G}\right]^{\frac{1}{3}}\left[a_t d_p\right]^{-2} \quad \text{(Eq 2-12)}$$

$$k_L\left[\frac{\rho_L}{\mu_L g}\right]^{\frac{1}{3}} = 0.0051\left[\frac{L_W^*}{a_W\mu_L}\right]^{\frac{2}{3}}\left[\frac{\mu_L}{\rho_L D_L}\right]^{-0.5}\left[a_t d_p\right]^{0.4} \quad \text{(Eq 2-13)}$$

where:

V^*_W = air mass flow rate per unit column cross-sectional area
ρ_G = density of air
μ_G = viscosity of air
D_G = gas-phase binary diffusivity
D_L = liquid-phase binary diffusivity
d_p = packing size

11. The overall mass transfer K_L is then computed.

$$\frac{1}{K_L} = \frac{1}{K_L} + \frac{1}{k_g K_H^{'}} \quad \text{(Eq 2-14)}$$

12. Compute the height of a transfer unit (HTU), assuming $a = a_w$.

$$HTU = \frac{L_W^*}{\rho_L K_L a} \quad \text{(Eq 2-15)}$$

13. The height of the tower required is then computed. A safety factor can be added to the height of packing, if required.

$$Z = HTU \times NTU \quad \text{(Eq 2-16)}$$

14. Head losses through the packing itself can be estimated from manufacturers' literature. The pressure drop across the demister packing support plate, duct work, and tower inlet and outlet is calculated.

$$\Delta P = k_p \left(\frac{Q_g}{A_x} \right)^2 \quad \text{(Eq 2-17)}$$

where:

A_x = the cross-sectional area of the packet tower aerator

15. The total horsepower requirement can be estimated from

$$AirHP = \frac{144 Q_g (p_1 - p_2)}{33{,}000(E)} \quad \text{(Eq 2-18)}$$

where an assumed fan and motor efficiency of 50 and 70 percent, respectively, can be used to yield an overall efficiency of 35 percent.

The tower air outlet in PTA systems must be designed to prevent noise from becoming a nuisance to neighbors. Sound regulations vary by site and zoning, but often the design must meet a decibel level at the property line. Air quality permitting may also be required depending on the contaminants removed and the location of the facility (ASCE/AWWA 2012).

Residuals from aeration systems are off-gases. In some applications, the off-gas must be collected and contaminants removed prior to discharge to the atmosphere. GAC canisters are most commonly used to collect and treat off-gases, with heating of the air flow to maximize efficiency and GAC life. Specific requirements for off-gas treatment from aeration facilities vary by location and with air quality standards.

Vacuum Degassers

A vacuum gas transfer system is usually executed as a tower aerator filled with a packing material in which the pressure is lowered by a vacuum pump. Membrane degassers are also used, where a vacuum is placed across a membrane and the gas is pulled from the water through the membrane (Macklin 2019).

Due to the vacuum pump, gas is removed from the water, resulting in lower gas concentrations. Because the gas concentrations in the water are lower than in the atmosphere, the saturation concentrations in the water are also lower. Because of the low saturation concentrations, it is possible to remove higher levels of gas from the water than is possible under atmospheric conditions. This makes a vacuum gas transfer system ideal for removing dissolved nitrogen and oxygen from the water and is frequently applied before the denitrification process. The efficiency of the vacuum gas transfer system depends on the vacuum pressure that is maintained in the treatment unit. The primary drawback of this technology is a higher energy use compared with packed tower aeration systems.

Staged Diffused Bubble Systems

There are two basic variations of bubble equipment as defined by the reactor and air introduction. Both of these designs use completely mixed (aerated) stages in series flow to approach a plug-flow efficiency. A completely mixed reactor is relatively inefficient compared to a plug-flow reactor because the mixing allows some portion of the influent to exit the reactor with minimal detention time in which treatment may occur. Therefore, the completely mixed stages must be placed in series to avoid back mixing and to achieve high efficiency. In this way, a plug-flow efficiency is closely approached if stages are greater than 3–6, depending on the contaminant and other factors. These systems are designed to maintain a relatively low profile compared to packed towers.

Diffused bubble systems almost exclusively use large orifices for bubble introduction to minimize maintenance. Air intensities are very high to minimize the equipment footprint, and this requires a high air velocity and a specific head loss at each orifice to maintain even air distribution without unnecessarily high pressure drop. For this type of equipment, a fine air bubble is not possible. Further, much lower maintenance is realized using coarse bubbles. Manufacturers simply design around any loss of transfer due to larger bubble size (Lowry and Lowry 1988).

The other important design parameter for these systems is depth. In testing many volatile contaminants during the 1970s and 1980s, it was learned that excessive depth was not needed for efficient bubble systems. Bubble rise depth has to be a balance of two factors: transfer efficiency and blower head (power required). Data show that there are a wide range of optimum depths when balancing these two requirements, based on the constituent being removed. For example, for the very poorly stripped methyl tertiary-butyl ether (MTBE), there is no significant advantage (transfer) beyond a depth of several inches. Transfer is minimal for this low Henry's constant VOC and it is transferred into an air bubble very rapidly. On the other extreme, carbon dioxide and radon transfer benefits from increasing depth up to several feet. The most important aspect of water depth is that several feet of depth is ample for optimum removal, and deeper water only results in an unnecessary increase in capital and operating cost associated with higher-pressure air blowers. This statement is valid for all applications of interest in drinking water treatment and for coarse bubbles.

The key to the design of this type of diffused bubble equipment is (1) limited depth and (2) coarse bubbles. Both of these parameters are fixed by the manufacturers of the equipment. Bubble aeration equipment that is designed using deep basins and finer bubble orifices are inherently expensive and less effective in drinking water applications (Lowry and Lowry 1988).

Vertically stacked bubble trays. This equipment uses vertically stacked successive shallow-depth trays with the water flowing downward by gravity and the air flowing in a counter-current, upward direction. A common name for this equipment is shallow tray air strippers. Depending upon the number of trays, each with 1–2 in. of water depth, the total static water depth varies from a few inches to tens of inches. Similar to packed towers, more height (trays) yields increased removal and increased blower pressure. Trays have holes for air flow (coarse bubbles), and downcomer pipes for water flow from tray to tray. An advantage of this design over packed towers is lower vertical height. Ease of maintenance is also reported to be an advantage, as trays are designed to be unstacked and packing is avoided (Ahmed et al. 2014).

Hundreds of these low-profile shallow tray air strippers have been used in pump-and-treat and drinking water applications for the removal of VOCs, radon, methane, hydrogen sulfides, and carbon dioxide.

General mathematical modeling for this type of equipment was described in 1958 in the *Handbook of Chemical Engineering*. However, each manufacturer has proprietary empirical design data and modeling available for its specific equipment.

A schematic and a photo of a stacked tray unit are shown in Figures 2-3 and 2-4, respectively.

Horizontal multi-staged bubble systems. Multi-staged bubble air stripping (MSBAS) and aeration have been used in drinking water treatment for 30 years. This type of system employs a vessel that is divided into stages using flow-through dividers and aerators in each stage. Water flows end to end, and air flows equally into each stage. This system was first applied to drinking water for very small water systems requiring extremely high removal (>99.999%) of VOCs in gasoline-contaminated wells (Lowry and Lowry 1984, Lowry and Lowry 1988) in the early 1980s. It has been used in larger systems for the removal of radon, carbon dioxide, VOCs, sulfides, methane, and nitrogen trichloride and for the addition of oxygen in the oxidation of iron (Lowry et al. 1984, Dixon et al. 1991, Schock et al. 1999). Several hundred systems are treating drinking water supplies. Recently, USEPA commissioned a study to define the performance of this system for the removal of 13 VOCs being considered for future regulation of carcinogenic VOCs (Lowry 2013).

Given a specific contaminant (Henry's constant) and water temperature, the basic design parameters for the process are

- air:water (A:W) ratio
- depth of bubble rise
- size of bubbles
- number of stages

The stages in the process are what create an efficiency far beyond what a single completely mixed vessel would produce. As the number of stages increases, the process

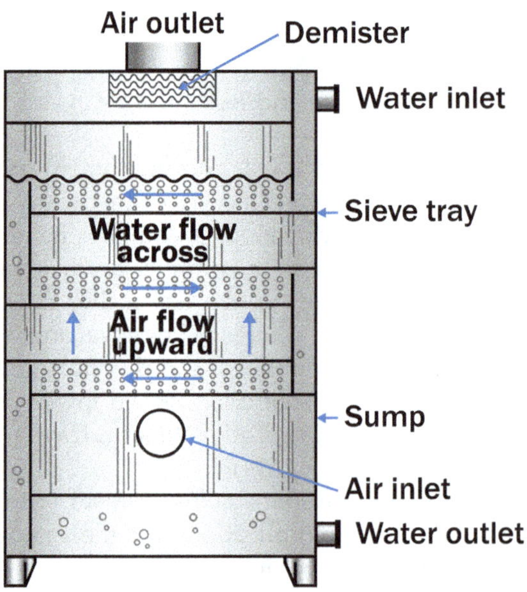

Courtesy of Carbonair.

Figure 2-3 Schematic of a stacked tray unit

Courtesy of Milt Larsen.

Figure 2-4 A stacked tray unit

efficiency approaches that of a theoretical plug-flow process. The two processes are shown below in Eq 2-19 and Eq 2-20.

Completely mixed stages in series:

$$C/C_o = [1/(1+kt)]^m \quad \text{(Eq 2-19)}$$

where:

C = concentration at steady state
C_o = inlet concentration
k = the first-order rate constant for a single stage, time^{-1} (directly proportional to air intensity, AI)
t = detention time, volume/flow, of a single stage
m = number of stages in series

Plug-flow:

$$C_t/C_o = e^{-kt} \quad \text{(Eq 2-20)}$$

where:

C_t = concentration at time t

Air is generally introduced to each stage at a high rate to minimize the equipment footprint. The Reynold's number for this type of air flow at an orifice is far greater than 5,000; therefore, the air exits the orifice as a continuous airstream and the bubbles do not form at the orifice. While the bubble size is influenced and proportional to orifice size, it is generally much larger, and the actual bubble size range and distribution are determined by the hydrodynamics of the mixing intensity in the stage.

It is important to note that the detention time in a completely mixed diffused bubble reactor is not important, as long as a specific A:W ratio is achieved. The first-order rate constant is directly proportional to the air intensity, all other factors being equal. For a fixed depth and aerator device, a doubling of the air intensity is a doubling of the rate constant for the range of practical application of these processes. The rate constant is an empirical quantity that is estimated using either a continuous pilot vessel or a batch reactor. As long as the depth, bubble size, and temperature are representative, then this constant is much more accurate than using a general theoretical model as described by others (AWWA 1999). The rate is exactly first-order ($r^2 \approx 0.99$), and design removal prediction is very accurate. Lowry has described this process in several publications, including those previously cited in this section (Lowry et al. 1984, Lowry 2013).

A schematic of a multi-stage bubble air stripper is shown in Figure 2-5. Example removal relations for carbon dioxide are shown in Figures 2-6 and 2-7. Figure 2-6 shows the first order relation for a single stage, and Figure 2-7 shows the resulting removal relation by application of the model for removal by completely mixed stages in series.

Summary

All of the types of gas transfer equipment described previously have their place in drinking water treatment. While packed towers and cascade type aerators are the most numerous throughout the United States, other types of equipment are applicable to many water supplies and are used routinely. Where high removal is required, packed towers and low-profile bubble systems are often required. Ease of maintenance, installed footprint, indoor installation, and overall height can be important factors in the determination of which equipment to use.

For special situations such as TTHM removal in relatively deep storage tanks, a surface or spray device can offer a solution without adding another step in treatment and/or additional pumping.

A photo of a multi-stage bubble system is shown in Figure 2-8.

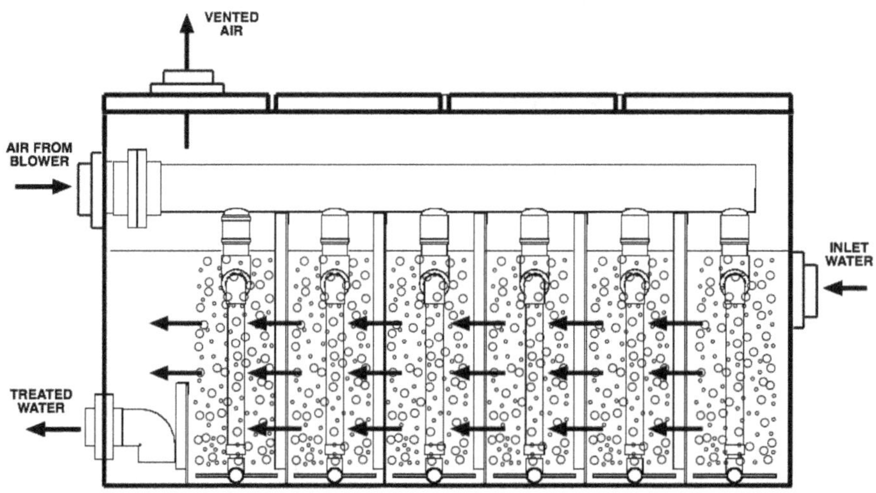

Courtesy of Lowry Aeration Systems, Inc.

Figure 2-5 Schematic of a multi-staged bubble air stripper

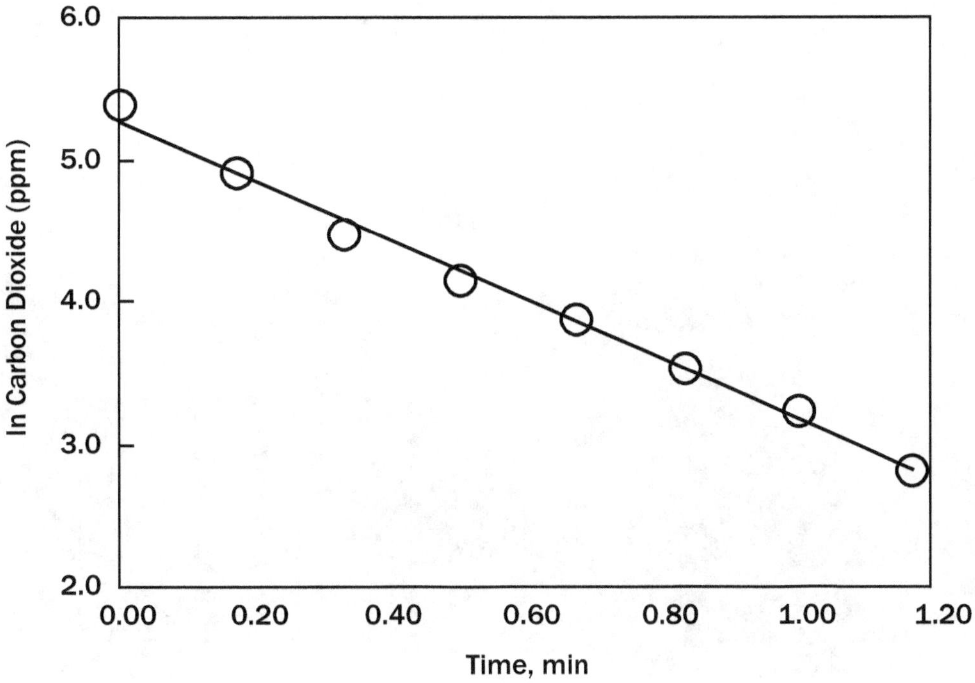

Courtesy of Lowry Aeration Systems, Inc.

Figure 2-6 First-order removal of carbon dioxide in batch test (k equals the slope of the fitted line)

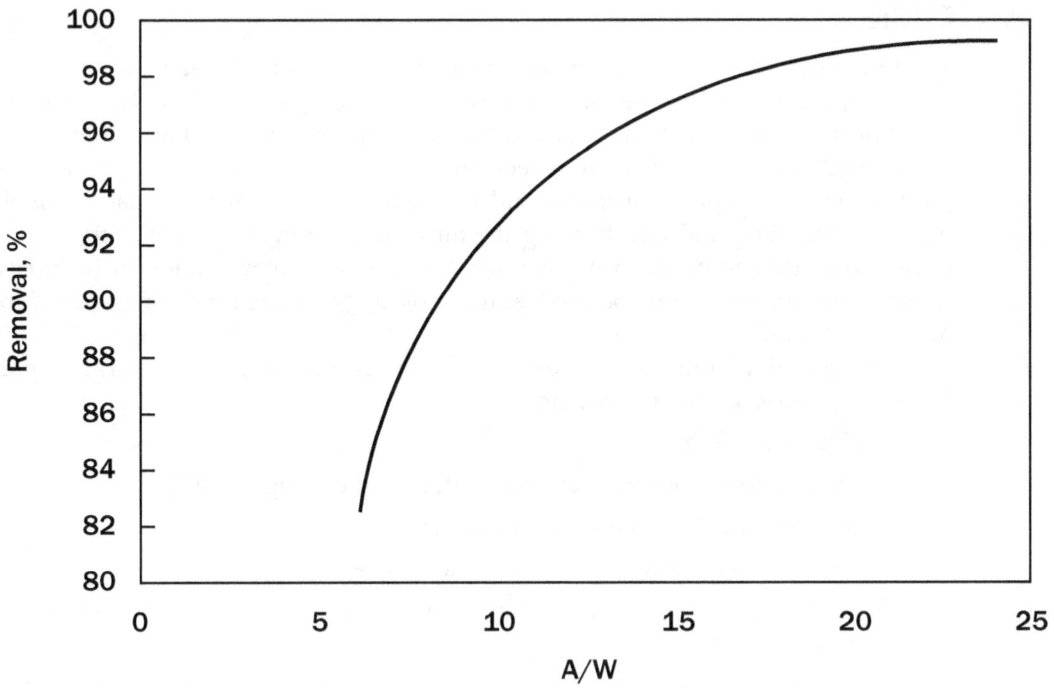

Courtesy of Lowry Aeration Systems, Inc.
A/W—air:water ratio

Figure 2-7 Removal of carbon dioxide in six-stage bubble reactor

Figure 2-8 Multi-stage bubble system for the removal of carbon dioxide at 500 gpm (Searsport, Maine)

BLOWER SELECTION

The terms *fan, blower,* and *compressor* have been used interchangeably over the years in various applications. In general, these terms are typically used for devices that span a wide range of pressure, with axial fans being on the low-pressure end and compressors being on the high-pressure end of the spectrum. These devices are negative-displacement or positive-displacement. Centrifugal and regenerative blowers are negative-displacement, where accelerating and decelerating the air achieve compression. Positive-displacement devices directly compress air by reciprocating pistons, rotary vanes, or turbines. For this chapter, the term *blower* will be used, as it is most appropriate for the majority of equipment being considered.

The selection of the air blower is based on several design and operating parameters. The blower must be able to provide

- clean, oil-free air,
- the required amount of air as dictated by the design A:W,
- the required design head for the application, and
- a noise level below an acceptable design value.

In addition to meeting design and operating parameters, the blower should be selected to minimize

- the energy required for the specific air stripping process that is used and
- the maintenance required for long-term operation.

Types of Blowers

The blower categories that are suitable for drinking water applications are axial fans, single and multiple stage centrifugal blowers, regenerative blowers, and positive displacement blowers. The range of blower types is a result of air volume and pressure requirements for a given application. Fans produce airflow at very low pressures and are appropriate for exchange of the air headspace in an enclosed storage tank using a spray device to air strip contaminants. At the opposite end of the application spectrum is the positive displacement blower. This type of blower is used for applications requiring relatively high air pressures to overcome the static pressure head of deep storage tanks. Between these two extremes are the blowers that are used most often for air-stripping applications involving packed towers and low-profile bubble and tray processes. These blowers are used to provide air at pressures in a range of a few inches of WC to several psi. The same blowers can also be applied to produce a vacuum in the process, as in specially designed packed towers where enhanced mass transfer may result from a negative pressure.

Each of the blower types described previously is shown in Figures 2-9 through 2-12.

The types of blowers most suited for aeration and air stripping are regenerative (lower flow rates) and centrifugal. These blowers provide air at the most reasonable power cost. Excess water depth or packing losses should be avoided, to minimize the blower pressure that is required. The theoretical power a blower uses is related to head by:

$$AHP = (Q \times P) \div 6{,}356 \qquad \text{(Eq 2-21)}$$

where:

AHP = air horsepower
Q = volumetric flow rate (SCFM)
P = pressure (inches of water gauge) or resistance

Source: Cincinnati Fan

Figure 2-9 An axial fan

Courtesy of Lowry Systems, Inc.

Figure 2-10 A centrifugal blower

Courtesy of Lowry Aeration Systems, Inc.

Figure 2-11 A regenerative blower

Diffused bubble vessels or trays should not require a design using a positive-displacement blower but instead typically use a centrifugal or regenerative blower. Water depth and blower head do not need to be greater than 3–4 ft for aeration or air stripping applications, based on fundamental transfer kinetics and efficient design. As previously explained, detention time, per se, is not important to removal with diffused bubble systems. Water

Courtesy of Lowry Aeration Systems, Inc.

Figure 2-12 A positive displacement-type blower

depth greater than 3–4 ft is not beneficial from a treatment standpoint and is extremely costly on an energy and capital cost basis.

Deep tank aeration or air stripping is a wasteful practice and should not be used. While it may be tempting to use an existing deep tank, it is always far more cost-effective to use a smaller and more cost-efficient design. For example, a tank that is 15 ft deep would require approximately 5 times the power of a vessel with a depth of 3 ft, and the blower type would be significantly more costly to operate and maintain. Higher water depths and blower pressures are not required for bubble systems and should be avoided.

Since packed towers have relatively low head for airflow, they typically use a centrifugal blower or fan. Packed towers normally require less blower head than bubble systems because they do not have a static water depth to overcome.

SCFM, ACFM, and ICFM

Air is compressible and its movement is typically expressed as a volume rate rather than a mass flow rate. In the US, the blower industry traditionally uses units of standard cubic feet per minute (SCFM), actual cubic feet per minute (ACFM), and inlet cubic feet per minute (ICFM). SCFM is the flow of air in cubic feet per minute at standard temperature and pressure. It must be noted that various temperatures and pressures have been referred to as "standard," depending on manufacturer and industrial application. The Compressed Air & Gas Institute (CAGI) and the American Society of Mechanical Engineers (ASME) standards state that standard air is at a temperature of 68°F, a barometric pressure of 29.92 (sea level), and a relative humidity of 36 percent. The density of standard air is 0.075 pounds per cubic foot. ACFM is the actual cubic feet per minute of air. ICFM is the inlet cubic feet per minute of air and is equivalent to the ACFM.

The ACFM is the actual volume of air moving at any temperature, pressure, and relative humidity, and is related to SCFM as follows:

$$SCFM = ACFM\left(\frac{T_1 + 460}{T_2 + 460}\right)\left(\frac{P_{atm}}{14.7}\right) \quad \text{(Eq 2-22)}$$

where:

T_1 = standard temperature, °F
T_2 = temperature of inlet air, °F
P_{atm} = atmospheric pressure at blower altitude, psi

Note that temperature and altitude are the main influences on converting ACFM to SCFM. It is common to disregard the correction for relative humidity because it is not a significant factor in most applications.

Once a blower is selected for the operating point at a given altitude, the changes in SCFM are almost entirely related to the inlet air temperature. The coldest seasonal temperature (highest air density) dictates the maximum power required (motor hp) and the warmest seasonal temperature (lowest air density) dictates the minimum pressure that the blower can generate at maximum speed (60 Hz). Both conditions must be checked to ensure the motor is adequate for the load and that the blower airflow will not drop below the surge point in the case of a centrifugal blower.

Affinity Laws

Blower speed influences the airflow, head produced, and the power required. The affinity laws, sometimes referred to as fan laws, can be used to correct a blower performance curve for speed change from N_1 to N_2, where N = speed in rpm or Hz. This applies to a blower operating on a variable-frequency drive (VFD). The affinity laws are as follow:

$$\left(\frac{Q_1}{Q_2}\right) = \left(\frac{N_1}{N_2}\right) \quad \text{(Eq 2-23a)}$$

$$\left(\frac{H_1}{H_2}\right) = \left(\frac{N_1}{N_2}\right)^2 \quad \text{(Eq 2-23b)}$$

$$\left(\frac{P_1}{P_2}\right) = \left(\frac{N_1}{N_2}\right)^3 \quad \text{(Eq 2-23c)}$$

where:

Q_1 = blower flow at speed N_1
Q_2 = blower flow at speed N_2
H_1 = blower head at speed N_1
H_2 = blower head at speed N_2
P_1 = blower power at speed N_1
P_2 = blower power at speed N_2

Minimum Speed, Airflow, and Blower Surge

When the speed of the blower is reduced, there are definite limits related to the head produced so that the blower pressure does not fall below the static head. This applies to types of air strippers that must overcome a minimum head to produce airflow. Diffused bubble and stacked tray air strippers have a fixed static water head that must be overcome to produce bubbles, and to distribute bubbles uniformly. This head is slightly higher during operation if an overflow weir is involved and the maximum water depth must be considered when a blower is operated at reduced speeds. The blower must always be operated at a pressure that is sufficiently high to avoid a deadhead condition.

Centrifugal and axial compressors will surge when forward flow through the compressor can no longer be maintained, due to an increase in pressure across the compressor, and a momentary flow reversal occurs. Once a surge occurs, the reversal of flow reduces the discharge pressure or increases the suction pressure, thus allowing forward flow to resume again until the pressure reaches the surge point again. This surge cycle will continue until some change is made in the process or compressor conditions. Severe damage to the thrust bearings, seals, and the impeller can occur with a severe surge condition. A centrifugal blower must not be turned down to a flow rate that puts it into a surge condition. The surge airflow limit is related to the internal design of the specific machine.

The extent of the damage is particular to the specific design of the blower, but total failure can occur with acute surge. An example would be if the blower inlet filter or intake screen becomes partially or totally blocked for period of time. When a blower is selected, the manufacturer normally provides the blower curve with minimum flow shown for each speed being considered. The minimum flow rates take into account blower surge and include a safety factor.

Surge can be an issue with a centrifugal blower because it is routine to throttle a centrifugal blower to produce a specific airflow and save power. Throttling a centrifugal blower reduces the power required, as power is related to the mass flow of air passing through the machine. The opposite is true for a regenerative blower, where throttling increases the power draw.

Blower surge is not a factor with a regenerative blower, and the lower limit on airflow is normally dictated by power draw and temperature rise. In the case of reduced motor speed for a regenerative blower, it is the required minimum airflow or the deadhead condition that dictates the minimum speed.

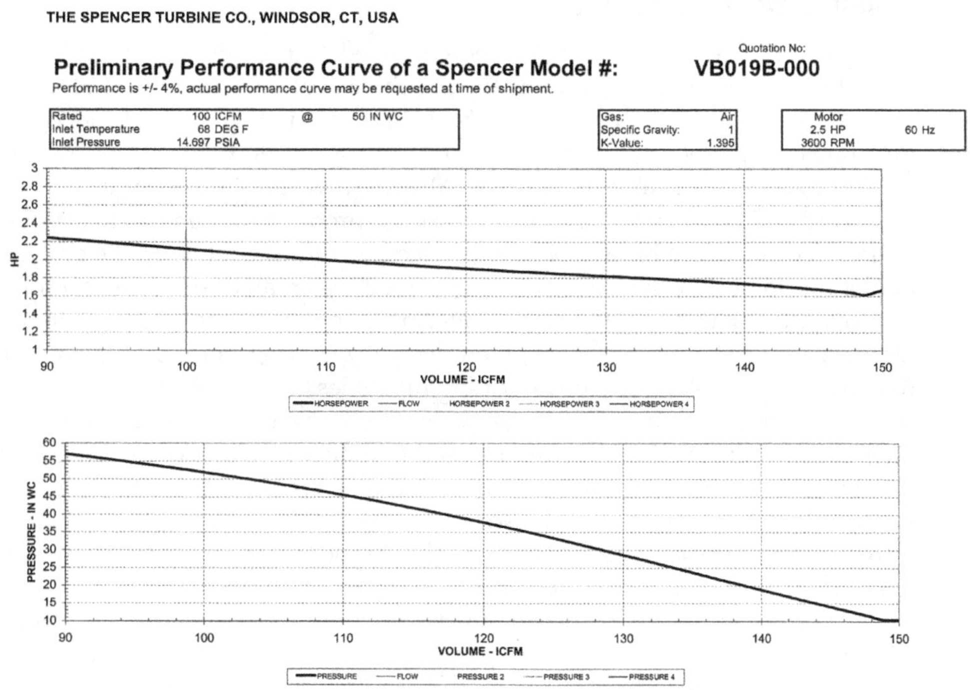

Courtesy of Lowry Aeration Systems, Inc.

Figure 2-13 Regenerative blower—scfm, pressure, power (Spencer Turbine Co., Windsor, Conn.)

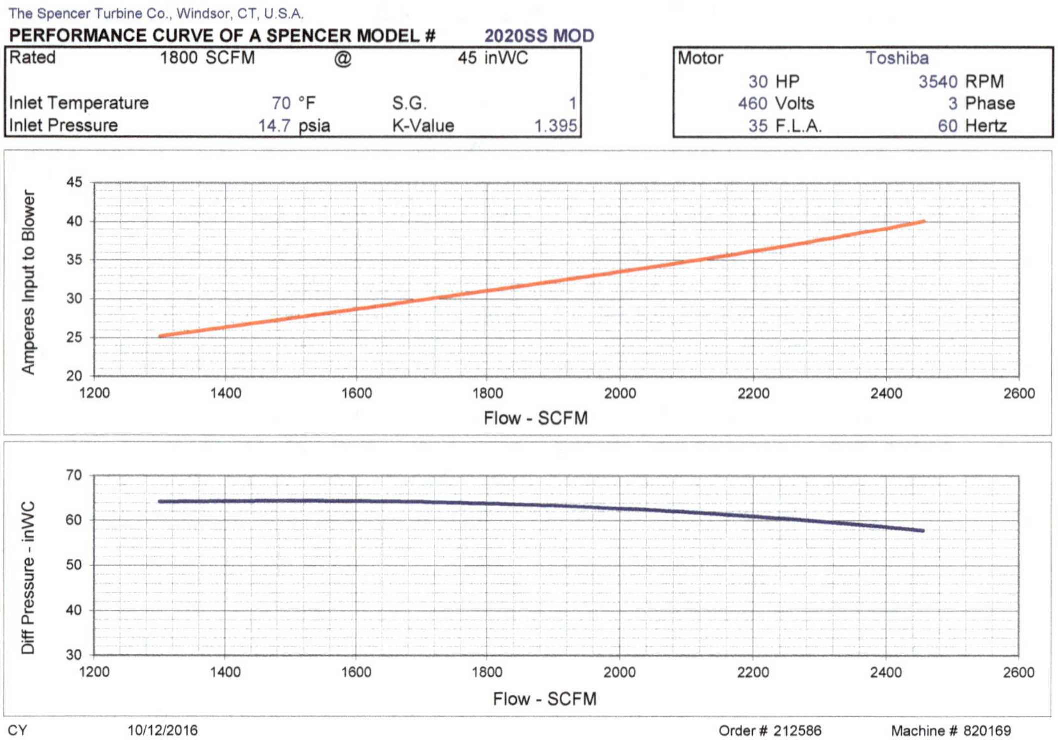

Courtesy of Lowy Aeration Systems, Inc.

Figure 2-14 Centrifugal blower curve—scfm, pressure, power (Spencer Turbine Co., Windsor, Conn.)

Required Operating Point

Any blower curve used must first be corrected to the site elevation for the project by correcting the pressure to account for the site elevation. The typical blower curves for a regenerative blower and a centrifugal blower, shown in Figures 2-13 and 2-14, respectively, illustrate the relationship for power required versus airflow. For the centrifugal blower, the minimum flow rate is terminated at 1,300 scfm to avoid blower surge.

A specific operating point is required to produce the required A:W for a given air-stripping design. The blower performance curve intersection with the system curve is the operating point. The system curve is obtained by an analysis of the total head required to move the air through the static water, packing, air filter, and the associated piping. Often, a blower is selected for a given design operating point that is equal to or greater than the required A:W. If the water flow rate is less than design, the blower will provide more air than is required. With a regenerative blower, the power is minimized without throttling the airflow. A VFD could be used to reduce power further and match the desired airflow more closely, but care must be taken to keep enough air flowing to provide even air distribution in any bubble air-stripping system. Regenerative blowers are typically for lower airflow applications involving higher static water requirements and therefore are not normally used on packed towers.

Centrifugal blowers have relatively flat performance curves. A selection is made for performance in excess of the operating point and throttling to the operating point is common. This ensures there is more than enough pressure available in the event of increased head loss in the system. Also, there are limits on the air intensity for the plan area of a given sized vessel or packed tower, and these often control the amount of air that can be input. Throttling the airflow with a butterfly valve on the discharge is common for single-stage lower head centrifugal blowers. These are the most common blowers for drinking water. In other industrial applications, where multi-staged centrifugal blowers produce much higher heads, it is preferable to have the throttling valve or vanes on the blower inlet. This results in lower air temperatures in the blower.

An alternative to throttling a centrifugal blower is to use a VFD. This is not a significant advantage with respect to power because a centrifugal blower draws less power simply by throttling with a valve; however, the VFD will avoid a high amperage draw upon startup and is sometimes useful where a reasonably constant A:W ratio is to be maintained for a widely varying water flow.

The total system curve is made up of the static water head of the bubble air stripper plus the intake and discharge piping losses for the range of airflow shown. It should be noted that the static water depth does not vary with temperature as do the frictional head losses in the piping. However, the overall effect of temperature is moderated because the kinematic viscosity of air is inversely related to temperature, so the water depth is the primary source of total system head. This would not be the case for a packed tower, where there is no static head of water to overcome. In any case, the highest head to overcome is always at the lowest ambient inlet air temperature. In practical application, the seasonal inlet temperature extremes (northern climes and higher elevations) will result in an increase of only ≈10–15 percent in frictional head loss, summer to winter.

The frictional head losses include the air intake, suction piping and fittings, the intake air filter, and the intake silencer, if present. The discharge losses include the blower discharge check valve, if present, the throttling butterfly valve, if used, the discharge piping and fittings, the orifice loss for the specific air-diffusing device used, and the losses in the vent piping. For a packed tower, there is the loss that is calculated for the packing at specific water and air loading conditions, etc., and all losses must be considered.

Once the system curve is calculated for the coldest seasonal temperature, the blower curve can be superimposed to arrive at the operating point. At minimum, the blower curve for the warmest seasonal temperature should be used to ensure that the blower pressure is adequate to meet the operating point desired. If a VFD is used, then various blower speeds can be used to determine the respective operating points.

A wide range of power required versus ACFM will be seen in blower curves for different inlet temperatures because the mass flow rate is varying. This wide range of power for different temperature disappears if SCFM is used to express airflow. Figure 2-14 illustrates that there is little variation in power versus air temperature if SCFM is used.

Motor Starter versus Variable-Frequency Drive

A standard magnetic motor starter/overload is most often used to drive the blower motor. In most cases, there is no need to vary the blower speed. The blower is selected, driven at 60 Hz, and there is excess air during the colder seasonal temperatures.

For designs that may require an A:W ratio to be maintained at a reasonably constant value over a wide range of water flows, a VFD can be useful. When a VFD is used, the minimum speed (airflow) must be determined to avoid blower surge in a centrifugal blower. A VFD is somewhat limited in the range of acceptable turndown because the air output for typical centrifugal blowers cannot be taken to less than 55–60 percent of the full speed output. This is to avoid surge. If the A:W ratio is to be maintained for the varying water flow conditions, then a mass airflow meter should be used, both to maintain the desired airflow with an analog signal and PI loop control and to maintain the airflow above the blower surge point.

Sound Level Control

Sound control can range from nothing to full blower enclosure. Axial fans are quiet and usually require no sound control measures. Many regenerative blowers require no sound control to meet Occupational Safety and Health Administration (OSHA) regulations (85 dB at 3 ft), as long as the air intake employs an air filter and the air discharge is underwater or into a packed tower. Inline silencers provide no real benefit in these types of installations.

Most centrifugal blowers require some degree of sound control to meet OSHA standards. With open intake and discharge, the sound level can be in excess of 120 dB and is excessive. Blowers are not operated in this manner, and sound levels are much lower. An air filter and silencer are normally on the intake, and the discharge is underwater or into a packed tower. Even with the intake and discharge connected in this manner, the blower is often too noisy to be comfortable. A standard practice is to use sound-wrap jackets on the blower housing to reduce the blower noise. This type of installation is sufficient for noise control in buildings or outdoors, where operators are occasionally present. Most installations have this level of noise control.

For more noise-sensitive locations, such as office areas and adjacent buildings, it may be necessary to provide additional noise control. If desired, a full sound enclosure can be designed to fully control the noise level. In those cases, it is difficult to hear that a blower is operating, and there is essentially no blower noise in the room.

Most blower installations that are indoors have an outdoor blower intake and a vent discharge to the outdoors. There can be a noticeable sound that emanates from the blower intake or vent, especially if a restrictive insect/bird screen is installed at the open end of the piping. This sound can be effectively minimized by selecting a screen with an open area of 65 percent or greater and enlarging the overall area of the pipe termination fitting. This practice will normally solve an intake noise issue even in very quiet residential locations.

For outdoor air stripper installations, often involving a packed tower or tray system, there is no building to provide any sound control. Additional control measures may have to be included in the design for sites that are in quiet residential locations. Often the site includes sufficient buffer area so that sensitive areas are at the required distance to eliminate noise issues. The sound level falls off in proportion to distance and is reduced by 6 dB for every doubling of distance from the source.

The prediction of sound level is difficult because many factors influence the actual sound level that will occur in a specific location. Distance, building size and materials, number of blowers, etc., are important factors. Occasionally, some sound control measures may need to be added after the installation and startup. A very effective measure is to use special sound-absorbing wall coverings inside the building.

Air Filtration

Air filtration is very important. It provides protection from particulate contaminants in the intake air and some degree of sound attenuation. Typical air filters for blower applications provide removal efficiencies of greater than 99 percent for either 2 micron (paper) or 5 micron (polyester). Once a filter element is partially loaded, the efficiency is increased but not determined. Air treatment beyond particulate removal is possible but rarely required or used. High-efficiency particulate air (HEPA) filtration is available at an added expense, and these filters remove particles to 0.3 micron at an efficiency of 99.97 percent.

Air filter selection should be focused on low head loss. This is not as important with regenerative blowers, since they typically have a relatively steep head versus airflow relation. For centrifugal blowers, with flatter performance curves, the system head losses should be minimized. Air filters should be sized to give an initial (clean element) head loss of less than 1–3 in. water column (WC). The filter element should be cleaned or replaced after the filter head loss increases by 3 in. WC. These are general guidelines and each installation is specific. The air filter is typically monitored with a differential pressure device.

Air filters can be intake or inline and are installed on the intake side of the blower. Intake filters are installed on the intake end of the piping, usually outdoors, and inline filters are indoors near the blower. Inline filters are more costly but offer easier servicing.

For removal systems that are inside a building, it is common practice to take air from the outdoors and discharge it to the outdoors. This eliminates an energy loss for the building. In cases in which the treatment plant is small and operates a few hours per day, the intake air can be from inside the building. In all cases, the vent air from the process is discharged outdoors.

In special locations where some specific restrictions may be required for the vent air, various filtration equipment, mist elimination, etc., may be added. All additional head losses must be added and equipment must be sized so the air stripper is not significantly pressurized. Vent pressures should be kept to below 1 in. WC pressure. If vent pressure is greater, an inline booster blower may be required in the vent piping.

Blower Monitoring and Protection

Blowers are monitored for intake filter and discharge pressure. Both pressures can change over time. Intake pressure increases when the air filter element becomes loaded with particulates. Discharge pressure can increase due to packing or orifices becoming fouled or clogged. In general, the pressures are monitored with local display to make it convenient for the operator to view the condition of the air filter and discharge while the system is running. Often, the pressure monitoring is included in the plant instrumentation and automation. Analog transducers are used to remotely monitor, control, and alarm specific conditions. At a minimum, the alarm conditions include high air filter pressure and high blower discharge pressure. Blower pressures are monitored in inches of WC. With transducer signals, warnings for filter element servicing and higher than normal discharge pressure can alert the operator prior to an actual alarm condition.

Centrifugal blowers must be protected from the low airflow surge condition. This can be done with a current switch or transducer for blowers that operate at 60 Hz. The centrifugal blower will draw less amperage at lower flow rates and the relation between blower motor amperage and air flow, in SCFM, is nearly linear and, for practical purposes, independent of inlet air temperature. A minimum amperage setting can be used to guard against a low airflow surge condition. A mass airflow meter is an alternative to using amperage for surge protection. The meter provides a direct measure of SCFM for any temperature and can be set up with instrumentation to alarm and shut down the blower if it operates at a lower than acceptable airflow.

Temperature monitoring is normally not needed for regenerative and centrifugal blowers in low-pressure applications. Motors are protected with overload devices, and the air is not compressed to a high temperature in aeration and air stripping applications.

Blower Maintenance

Maintenance requirements vary with blower type. Axial fans, regenerative blowers, and centrifugal blowers do not have wearing surfaces between the blower impeller and the housing. These types of blowers typically have two bearings and the bearings are either sealed, requiring no routine lubrication, or they have grease fittings for lubrication on a routine schedule. Large centrifugal blowers often have continuous oiling systems with reservoirs that require periodic filling.

Bearing failure is normally gradual and higher than normal amperage and/or noise should occur prior to damage of the machine due to a worn bearing. The bearings can be easily replaced on smaller blowers, and there are no other wearing parts. It is not uncommon for a regenerative or centrifugal blower to last decades with proper maintenance.

Vent and Air Intake

Location, screening, weatherization, and energy loss are important factors to consider with the air intake and vent discharge. Normally the controlling factor for all four is winter operation in colder climates.

Locating the air intake to prevent any cross-connection with vent discharge air is critical. If the air intake is able to draw the discharged air, removal efficiency of the air stripper suffers because there is contaminant in the "clean" air used to remove the contaminant. A contaminated air supply can be measured in a loss of treatment efficiency, especially for high removal efficiency designs. If the intake air contains the contaminant, then there is a lower limit, based upon equilibrium, to removal of that contaminant. A natural example of this effect is with carbon dioxide, where the air always contains a natural low level of carbon dioxide. Using air for the removal of carbon dioxide from drinking water, as in corrosion control air stripping, cannot produce water having less than the equilibrium level of approximately 0.6–0.8 mg/L.

Locating. The location of the air intake and vent discharge should not be on the same building wall. If this cannot be avoided, the intake and vent should be directed away from each other. Vents through a building wall should not be directed against a hillside or into a fenced enclosure, as either condition may trap the contaminated air and result in locally elevated concentrations. Extreme examples of this practice, combined with poorly located air intake, have resulted in treatment efficiency drops in excess of 50%. Relocating the intake and or vent immediately corrects this problem. If the vent is placed through the building roof, it should be designed to direct air upward or outward, rather than downward toward an air intake. While contaminants that are air-stripped are never at a concentration that would come out (condense) of the bulk air, very cold air on a warm day can sink rapidly due to the differential density and lack of mixing.

In general, a minimal amount of consideration for the respective placement of the air intake and vent discharge will provide ample segregation of the air streams. Mixing and dissipation of the vent discharge are normally very thorough and rapid, so a distance of more than 10–15 feet separation is sufficient. However, in very large operations with large air volumes, more distance may be needed. A very effective practice is to elevate the exit point of the air vent.

Screening and cold climates. Insect or bird screens are used on air intakes and discharge vents. Air filters provide ample screening for the blower intake filters located at the end of the intake piping. Inline air filters also do the same, but often screening is used on the end of the intake piping to avoid buildup of insects, leaves, or other debris in the piping between the outdoors and the inline filter housing. Similar screens are often used on air discharge vents. Insects or birds could enter the vent during times when the plant is not operating.

Screening of intakes and vents should be done with stainless steel mesh screen with a minimum open area of 65 percent. Care must be taken to not specify a screen that is too restrictive due to small mesh size or a low open area. The same mesh size can have a large variation in open area due to varying wire diameter in the construction of the screen. A good compromise for opening size and restriction is a mesh size of 4×4 or 6×6, with an open area in the range of 65 percent.

Consideration should be given to using the next diameter pipe size on the end of the intake and vent piping. This will ensure that the screened end will not be a restriction with respect to airflow.

Cold climates pose additional requirements for air intake and vent discharge. In general, all piping exposed to the outdoors in cold climates should be at least 4-in. diameter. This is normally the case anyway for all but the smallest air stripping operations. Particular attention must be given to icing and snow conditions to avoid icing of too restrictive screens and the suction of snow into the blower piping/filter. In general, the amount of exposed piping should be minimized.

Hazardous contaminant consideration. Special attention must be taken when removing specific contaminants that can pose hazards. These include any health-related contaminants, or ones that pose a risk of explosion. Specifically, sulfides, methane, and nitrogen trichloride (trichloramine) are contaminants that can create a serious risk. In general, VOCs do not pose any real risk for discharge due to the extremely low levels present in the water supply, the high A:W used to remove them to low levels, and the rapid dissipation into the surrounding air.

Sulfides. Air stripping operations can have dangerous levels of hydrogen sulfide in the vent discharge, prior to dissipation into the surrounding air. Calculations that consider the level of sulfides, removal efficiency, and the A:W must be made to ensure that dangerous levels of hydrogen sulfide are not able to come into contact with people. The resulting concentration of hydrogen sulfide must be compared to the acceptable regulated exposure levels to ensure worker safety. If a potentially dangerous condition exists near the vent discharge or inside a building, then hydrogen sulfide monitoring and alarming would be included in the air stripper design. Complaints of typical rotten egg odor are often associated with the stripping of sulfides.

In addition to a health hazard, hydrogen sulfide is extremely corrosive to moist metallic surfaces above the waterline. These materials, including stainless steels, are particularly vulnerable to rapid pitting corrosion. Materials subject to corrosion should be avoided or-coated with epoxy paint.

Methane. Explosions have occurred in homes and in storage tanks in drinking water systems that contain methane. Natural methane in groundwater is not rare in certain parts of the US and elsewhere. The solubility limit of methane in water at 20° C is approximately 23 mg/L. There are water supplies that have this level of methane present in the well water as it comes out of the ground. Since methane is very volatile, the lower explosive limit for methane (LEL = 5%) can easily be exceeded in a confined headspace or poorly ventilated bathroom. Certainly, an air stripper treating this type of water supply must take into account the potential risk and make certain that untreated water does not sit idle in an enclosed tank. During operation, the A:W is always more than sufficient to take the methane concentration in the air discharge to well below any danger level; however, for a diffused bubble air stripper the blower should be controlled to stay operating for a period of ≈2 min after the well water flow is stopped. This ensures that the remaining methane present in the air stripper is removed to a low level. Explosion-proof motors are recommended if there is a possibly explosive atmosphere in the vicinity of any motors or controls. In some locations, air treatment devices are used on the discharge vent of air strippers. In those cases, booster blowers would be required to avoid high air pressure inside the air stripper. The air treatment devices also need to be designed for minimal head loss.

Nitrogen trichloride. Both chlorine gas and nitrogen trichloride are strippable from water. Chlorine gas is not present at a significant concentration (≈ <10 ppt) at drinking water pH levels, since it forms hypochlorous acid and hypochlorite ion. However, with breakpoint chlorination in chloramination operations where the Cl_2:N is higher than 5, significant nitrogen trichloride is formed even in the pH range of 7.0–7.5. At lower pH, more is formed.

Nitrogen trichloride is extremely corrosive to unprotected steel and other materials, including stainless steels and elastomers. In addition, it causes asthma, especially in young children. While it would be a relatively rare occurrence, it is possible to have conditions where the vent from an air stripper would contain very high levels of nitrogen trichloride, in comparison to levels that are known to cause asthma, burning mucous membranes, etc. In these specific cases, care should be taken to locate the air discharge vent to minimize exposure. Complaints of odor are possible with air stripping of nitrogen trichloride.

Radon. The risk from radon is long-term inhalation exposure. While radon does not pose a significant health hazard in the short term, any air stripping operation should consider the proximity of the vent discharge to workers. Radon occurs at some level in almost all groundwater supplies. The US air limit for indoor air is 4 pCi/L and many air stripping operations removing radon will have a vent discharge that is greater than this level. Actual radon levels in air discharge vents are as high as 10,000 pCi/L in extreme cases. These are very high radon wells. Typical wells with 1,000 pCi/L in the well water may have vent concentrations of 100 pCi/L. Studies have shown that radon is dissipated very effectively in a short distance from the vent discharge, so it is relatively easy to minimize the potential for exposure (Narasimhan et al. 2005).

REFERENCES

Ahmed, A.T., et al. 2014. Removal of Volatile Organic Contaminants via Low Profile Aeration Technology. Denver, CO: USEPA, Water Environment Research Foundation, and Water Research Foundation.

Appropedia, n.d. Original:Slow sand filtration water treatment plants. http://www.appropedia.org/Original:Slow_sand_filtration_water_treatment_plants. Accessed Oct. 12, 2020.

ASCE/AWWA. 2012 (5th ed.). *Water Treatment Plant Design*. New York: McGraw-Hill, Inc.

Ashworth, R.A., G.B. Howe, M.E. Mullins, and T.E. Rogers. 1988. Air-Water Partitioning Coefficients of Organics in Dilute Aqueous Solutions. *Journal of Hazardous Materials*, 18(1):25–36.

AWWA (American Water Works Association). 2010 (6th ed.). *Water Quality and Treatment*. New York: McGraw-Hill, Inc.

Danckwerts, P.S.V. 1970. *Gas-Liquid Reactions*. New York: McGraw-Hill.

Delft University of Technology. 2019. Water Treatment Open Coursework. https://ocw.tudelft.nl/wp-content/uploads/Aeration-and-gas-stripping-1.pdf.

Dixon, Kevin L. 1991. Evaluating Aeration Technology for Radon Removal. *Journal AWWA*, 83:4.

Dyksen, J. E. (3rd ed.). Chapter 5: Aeration and Air Stripping, Water Treatment Plant Design. American Water Works Association. New York: McGraw-Hill, Inc.

General Electric. 1997–2009. *Handbook of Industrial Water Treatment*", General Electric Company: Fairfield, Conn. http://www.gewater.com/handbook/index.jsp (accessed Oct. 12, 2013).

Gmehling, J., U. Onken, and W. Arit, 1979. Vapor-Liquid Equilibrium Collection (continuing series), DECHEMA, Frankfurt.

Gossett, J. M. 1987. "Measurement of Henry's law constant for C1 and C2 chlorinated hydrocarbons" Environ Sci. and Technol. 21:202–208.

Hand, D.W., D.R. Hokanson, and J.C. Crittenden. 2011 (6th ed.). Chapter 5: "Air Stripping and Aeration." In: Water Quality and Treatment: A Handbook of Community Water Supplies. American Water Works Association. New York: McGraw-Hill, Inc.

Hirschfelder, J.O.; Bird, R.B.; and Spotz, E.L. The Transport Properties of Gases and Gaseous Mixtures. *Chemical Reviews II*, 44(1):205–231.

Kavanaugh, M. and R. R. Trussell. 1980. "Design of Aeration Towers to Strip Volatile Contaminants from Drinking Water." *Journal AWWA*, 24(1):62.

Landolt-Bornstein database, II/8: Magnetic Properties of Coordination and Organometallic Transition Metal Compounds (1976).

Leighton, D.T., Jr. and J.M. Calo. 1981. Distribution coefficients of chlorinated hydrocarbons in dilute air-water systems for ground-water contamination applications. *Journal of Chemical Engineering Data*, 26(4): 382–385.

Lowry, Jerry D. 2013. Multi-stage Bubble Air Stripping Tests, Contract No. EP-C-11-006, NMRMRL-STD/LRPCD. Cincinnati, OH: USEPA.

Lowry, J.D., et al. 1984. "GAC Adsorption and Diffused Aeration for the Removal of Radon from Water Supplies," Proceedings of the National Conference of the American Water Works Association, Dallas, TX, June 1984.

Lowry, J.D. and S.B. Lowry. 1984. "Restoration of Gasoline-Contaminated Household Water Supplies to Drinking Water Quality," Proceedings of the Eastern Regional Ground Water Conference, Portland, ME, July 1984.

Lowry, J.D. and S.B. Lowry. 1988. "Removal of Petroleum Hydrocarbons and MTBE from Water by a Multi-Staged POE Aeration System," in Proceedings of Focus on Groundwater Conference, National Water Well Association, Stamford, CT, September 1988.

Lowry, J.D. "Corrosion Control with Air Stripping," Proceedings of the 2010 Inorganic Contaminants Workshop, American Water Works Association, Denver, CO, February 28–March 2.

Lowry, J.D. "Removal of Sulfides Using Carbon Dioxide and Multi-Staged Bubble Air Stripping," Proceedings of the 2010 Inorganic Contaminants Workshop, American Water Works Association, Denver, CO, February 28–March 2.

Mackay, D., W. Y. Shiu, and R.P. Sutherland. 1979. Determination of air-water Henry's law constants for hydrophobic pollutants. *Environmental Science and Technology*, 13(3): 333–337.

Macklin, S.H.; Haas, W.E.; and Miller, W.S. Suez Water Technologies Technical Paper, Carbon dioxide and dissolved oxygen removal from makeup water by gas transfer membranes.

Montgomery, J.M. 1985. *Water Treatment Principles and Design*. New York: John Wiley and Sons.

Odell, Lee H. 2010. *Treatment Technologies for Groundwater*. Denver, CO: AWWA.

Onda, K.; Takeuchi, H.; and Okumoto, Y. 1968. Mass transfer coefficients between gas and liquid phases in packed columns. *Journal of Chemical Engineering of Japan*, 1:56–62.

Schock, M.R.; Lovejoy, T.R.; Holldber, J.; Lowry, J.D.; and J. Egan, 1999. "California's First Aeration Plants for Corrosion Control," Proceedings of the 1999 Annual AWWA WQTC, Tampa, FL, November 1999.

Sherwood, T.K., R. L. Pigford, and C.R. Wilke. 1975. *Mass Transfer*. New York: McGraw-Hill, Inc.

Sherwood, T.K. and F.A.L. Holloway. 1940. Peformance of packed towers – liquid film data for several packings. *Transactions of the American Institute of Chemical Engineers*, 36:39–70.

US Army Corps. 2001. Engineering and Design Air Stripping, Department of the Army, US Army Corps of Engineers, Washington DC, Design Guide No. 1110-1-3, October 31, 2001.

USEPA. 1989. Design Manual, Fine Pore Aeration Systems. EPA/625/1-89/023.

Wilke, C.R. and P. Chang. 1955. *Correlation of Diffusion Coefficients in Dilute Solutions, A.I. Ch.E. Journal*, June.

This page intentionally blank.

Chapter 3

Gas Addition

Gas addition in drinking water supplies is generally limited to introduction of a number of gases:

- Atmospheric air
- Oxygen
- Carbon dioxide
- Anhydrous ammonia
- Chlorine gas
- Ozone
- Sulfur dioxide

The focus of this chapter is mainly the introduction of atmospheric air and oxygen, although the other gases are discussed.

APPLICATIONS

Air is commonly used for iron oxidation with oxygen and to promote the establishment of manganese dioxide coatings to catalyze the oxidation of manganese. It is also used in biological water treatment processes to promote biological growth. Oxygen is sometimes used for these same purposes. Carbon dioxide is commonly used to adjust pH in lime softening, arsenic adsorption filters, and in the removal of sulfides. Chlorine gas is commonly used for primary disinfection and the addition of anhydrous ammonia to chlorinated water is a common method to produce chloramines. Ozone, mixed with oxygen, is used for ozone disinfection and oxidation. Sulfur dioxide gas is used to dechlorinate water supplies.

IRON AND MANGANESE OXIDATION

A common method for removing iron and manganese from water is aeration, followed by precipitation and filtration. Figure 3-1 shows a picture of an aerator. Aeration often raises the pH of the water, by stripping off excess carbon dioxide, and the higher pH is favorable to the kinetics of oxidation. Ferrous iron (Fe^{++}) is easily and quickly oxidized

Courtesy of Milt Larsen.

Figure 3-1 Packed tower air stripper under construction

in many groundwater supplies, but organically-complexed iron is not. Manganese is oxidized very slowly by oxygen in air without a manganese dioxide surface present. Oxidation rates increase with increasing pH, but often requires a stronger oxidant like chlorine or permanganate to be used. Aeration and chlorination have both been cited as agents to oxidize manganese, but without the manganese dioxide adsorption surface that coats granular media after several days or weeks of operation, nether is effective. Once the oxide coatings develop, oxygen and free chlorine are effective. In fact, some manganese oxidation will occur inside aeration equipment when the oxide coatings have become established. If the pH of the groundwater is below 7.0, even after aeration, chlorine may be required to fully oxidize iron. However, field experience has often demonstrated that air alone produces a superior iron floc for filtration as compared to chlorine. This is a good reason to use aeration for iron removal.

Different oxidants can be used to precipitate iron and manganese. It is important to consider the amount needed for each oxidant, the reaction time, and interferences with oxidation. The amount of chemical needed to oxidize iron and manganese is shown in Table 3-1, and the range of reaction times is shown in Table 3-2. Only 1 mg/L of oxygen is required to oxidize 7 mg/L of dissolved iron, and 1 mg/L of oxygen is required to oxidize 3.5 mg/L of manganese. Because such small amounts of oxygen are needed, even groundwater systems that do not chlorinate and have very low DO levels in the distribution system can expect oxidation and precipitation of iron and manganese.

Several compounds can interfere with oxidation including hydrogen sulfide, organic compounds, and ammonia. Ammonia exerts a 10:1 demand on chlorine, and organic compounds can affect both the amount of chemical demand and the reaction time by complexing with iron and manganese.

Table 3-1 Oxidant requirements for iron and manganese

Oxidant	Per mg/L of Mn	Per mg/L of Fe
Oxygen (from aeration)	0.29	0.14
Ozone	0.67	0.43
Chlorine	1.28	0.63
Potassium permanganate	1.92	0.94
Chlorine dioxide	2.4	1.2

Source: Odell 2010

Table 3-2 Oxidation reaction times for iron and manganese

Oxidant	Per mg/L of Mn	Per mg/L of Fe
Oxygen (from aeration)	80 min to 2 days, pH-dependent	<1 min to 1 h, pH-dependent
Ozone	<5 min	<2 min
Chlorine	15 min to 12 h, pH-dependent	<1 min to 1 h, pH-dependent
Potassium permanganate	<7 min	<5 min
Chlorine dioxide	<5 min	<5 min

Source: Odell 2010

BIOLOGICAL TREATMENT OF GROUNDWATER

Biological removal of iron can be accomplished by promoting growth of certain bacteria including *Gallionella, Leptothrix,* and *Sphearotilus* on a media surface. For biological removal of manganese, autotrophic bacteria require specific conditions for effective removal. The most important conditions for biotic uptake of iron are

- pH within the range of 5–9
- Change from reduced redox conditions to a slightly positive redox conditions of 200–320mV
- Oxygen levels up to 3 mg/L
- Carbon dioxide present in the water

These conditions can be readily accomplished through oxygen or air addition to many groundwater supplies.

Uptake of iron and manganese in biological removal systems can be quite rapid. Both iron and manganese require a period of startup before bacterial populations reach a density that makes removal effective. After periods of inactivity, a shorter startup time may also be required. When short-term shutdowns of a few hours occur, iron and manganese removal generally returns to normal after a few minutes. If shutdowns last for several weeks or months, the startup period may take several hours. Media should be kept wet during shutdowns to minimize the startup period.

Oxygen added to biological filters can be used for ammonia removal by promoting the growth of ammonia oxidizing bacteria which convert ammonia to nitrite and then to nitrate (Lytle et al. 2013).

DISINFECTION

Disinfection is required to inactivate pathogens in drinking water and wastewater treatment. Disinfectants are generally strong oxidants that are often naturally present in gaseous form. In order to introduce the gaseous disinfectants to the water, aeration is required. The main chemical disinfectants that are added to water in their gaseous state are

- chlorine
- chlorine dioxide
- ozone
- anhydrous ammonia

Gaseous Disinfection Agents

Chlorine. Chlorine gas, which is stored in pressurized containers as liquefied chlorine, can be introduced into water through either gas-to-solution systems or direct gas injection, which is not commonly used. The biggest issue of utilizing aeration for chlorine gas introduction is safety. Chlorine gas is extremely hazardous, so safe storage, handling, and use of the chlorine are key. Direct gas injection systems are mainly utilized when the gas storage location is in close proximity to the gas introduction location, so probability of a leak is low. The gas-to-solution systems are utilized when the gas storage location is further away from the gas introduction location. In the gas-to-solution systems, the chlorine is metered by a chlorinator as a gas, and then mixed with water in an injector/eductor. Figure 3-2 shows a picture of an injector system.

Chlorine will form trihalomethanes (THMs) and haloacetic acids (HAAs) when it reacts with organic matter in water. THMs and HAAs are regulated disinfection byproducts (DBPs), so chlorine is mainly used as a primary disinfectant, and its use as

Courtesy of Milt Larsen.

Figure 3-2 An injector system

a secondary disinfectant to maintain disinfectant residual is decreasing. For secondary disinfection, chlorine is often combined with ammonia to form chloramines, which have a reduced potential for the formation of THMs and HAAs. Chlorine can also be used as a pre-oxidant, where the chlorine reacts with contaminants prior to disinfection, but a residual is not left. Chlorine is often used as the secondary disinfectant. Many utilities use chlorine as part of their distribution system flushing and disinfection program, but chlorine gas is not normally used for these programs, as feeding chlorine gas at remote, normally residential locations is not safe. Sodium hypochlorite or calcium hypochlorite is generally used for these applications and can also be used in applications where it is deemed that storing and feeding chlorine gas is not safe.

Chlorine dioxide. Chlorine dioxide gas is generally produced on-site with the chlorine-chlorite process. Safe generation, handling, and use of chlorine dioxide are also key, since chlorine dioxide is hazardous. Feeding chlorine dioxide is generally accomplished by injection, similar to chlorine gas. Chlorine dioxide has residual byproducts of chlorite and chlorate. Both chlorite and chlorate are regulated disinfection byproducts, so the use of chlorine dioxide as a disinfectant is not common.

Ozone. Ozone gas is generated on-site, due to its instability. The ozone gas is generated by applying an electric field to oxidize oxygen (usually introduced in air, rather than pure oxygen) to form ozone. The ozonator will produce ozone gas and meter it. Ozone gas is then piped to a contactor where it is introduced to the water to be treated. The ozone is introduced either by side-stream injection, sparging or diffusers, which are discussed in Chapter 2 of this manual.

Ozone reacts with bromine in water to form bromate, which is a regulated disinfection byproduct. Ozone is not typically utilized as a primary disinfectant in drinking water treatment, because it does not impart a protective residual to the distribution system. However, it has been used as a pre-oxidant. Ozone is an attractive disinfectant for bottled water, since a residual is not required.

Anhydrous ammonia. Anhydrous ammonia is utilized in disinfection to react with chlorine in the water to form chloramines. Chloramines have the same regulated disinfection byproducts as free chlorine. Chloramines are attractive as secondary disinfectants, since they have a longer residual in distribution systems than chlorine. Anhydrous ammonia is stored in cylinders or pressurized tanks. Ammonia can be added to the water either through direct injection to the water being treated or through gas-to-solution methods. Direct injection involves an ammoniator, which will meter the anhydrous ammonia into the water to be treated under positive pressure. The gas-to-solution method is similar to chlorine gas, where an injector/eductor is used to introduce ammonia in solution to the water being treated.

Design of Gaseous Disinfection Systems

Design of Disinfectant Injector/Eductor Systems. Designs of injector/eductor systems are similar, no matter the gas-phase chemical being fed into the water to be treated. The main design element for these systems is determining how much of the gas needs to be fed.

When designing a system for chlorine or anhydrous ammonia, the amount of gas to be stored on-site is based upon the amount of the chemical needed to be fed under average conditions. Generally, utilities will have a 15- or 30-day supply of chemical on-site, with the actual feed cylinders/tanks on scales sized to supply the chemical for a few days. The gas storage and feed system are often designed with a redundant system, so that automatic switchover to the redundant system can occur, preventing a lack of disinfectant feed at times when the water treatment plant is not manned.

The amount of chemical to be fed depends on the demand for the chemical. For instance, chlorine demand is calculated by determining the amount of chlorine residual desired, the amount of chlorine required to obtain the contact time (CT) desired for inactivation of microorganisms, and the concentrations of other chemicals in the water that compete with microorganisms to react with chlorine, such as organic material, sulfur compounds, and ammonia, and calculating the stoichiometric amount of chlorine needed to reach the desired residual. Naturally occurring ammonia will not always result in monochloramine, so breakpoint chlorination is often desired if naturally occurring ammonia is present. However, depending on water quality, nitrogen trichloride can form with breakpoint chlorination (high Cl:N ratio) and this should be avoided. Nitrogen trichloride gas is volatile and extremely corrosive. It has been implicated in the premature failure of both polyethylene piping and unprotected mild and stainless steels (Lowry 2013, Rockaway et al. 2007, Reiber 1993, White 1992). Nitrogen trichloride can be removed by aeration, and this must be done to avoid downstream corrosion problems in the distribution system. The Henry's constant for nitrogen trichloride is similar to TCE and it can be stripped (Lowry 2013). A photo of stainless steel corrosion caused by nitrogen trichloride is shown in Figure 3-3.

The demand for the chemical is calculated based upon both the average and peak flow demand conditions. The average flow demand is calculated by the design engineer based on the population being served by the system for the design year. Generally, the peak flow demand is calculated based on a peaking factor that is applied to the average

Courtesy of Lowry Aeration Systems, Inc.

Figure 3-3 Photo of stainless steel corrosion due to nitrogen trichloride

flow demand. The peaking factor is often determined based on empirical data available for the system being served (or a nearby system) or based on requirements of the individual state environmental department or health department.

Injector/eductor systems are designed to feed the chemical for the entire range of minimum to maximum flow. The chemical is generally flow paced, where the chemical feed system calculates the demand based upon the flow of the water being treated. Control systems to dose the chemical both on flow pacing and residual control are also utilized.

Because of the hazardous nature of gas storage and delivery systems, compliance with Federal, State, and Regional regulatory entities must be evaluated for feasibility of installation. Additionally, if gas systems are implemented, a program educating and training operations staff on safe handling and use must be implemented.

Design of direct feed systems. Direct feed systems are designed similarly to injector/eductor systems, as explained in the above section.

Design of ozone contactors. Ozone contactors are designed to efficiently transfer the ozone into the water being treated. The ozone dose is calculated based upon the contact time desired to inactivate microorganisms as well as the amount of the chemicals that compete with microorganisms to react with ozone. Ozone is also fed based on flow pacing to maintain the contact time desired in the fixed volume of the contactor. The flow pacing is calculated by a programmable logic controller (PLC) based upon the ozone dose calculated by the design engineer to achieve inactivation by contact time and to provide enough ozone to overcome the chemical demand. The systems integrator will work with the design engineer to make sure that the proper equation is used for the flow pacing program.

Reduction/Dechlorination

Reduction is often required to quench an oxidizing chemical, as is the case with dechlorination. Dechlorination is used in drinking water treatment to reduce chlorine to prevent disinfection byproduct formation if chlorine must be overfed. Dechlorination is also required in some states for backwash water from water treatment processes prior to a discharge to an in-ground sand filter or a receiving stream. Sulfur dioxide gas is utilized to reduce chlorine, chlorine dioxide, and chloramines to chloride. Sodium bisulfite or sodium meta-bisulfite, which are both present in solution, are also commonly used reducing agents. Since sulfur dioxide is the only gas used to reduce disinfectants, it is the only reducing chemical discussed in this manual.

Sulfur dioxide gas is stored on-site in pressurized cylinders or tanks. Sulfur dioxide is fed into the water being treated similarly to chlorine, with a sulfonator to measure the flow of sulfur dioxide and an injector/eductor to add the gas into solution. The solution is then fed to the water being treated. Sulfur dioxide feed systems are designed to only add as much sulfur dioxide as is needed to reduce the chemical (generally chlorine). Over-feeding of sulfur dioxide is not desired, since the sulfur dioxide will then react with oxygen in the water to produce sulfates, which will decrease the pH of the water being treated. Sulfur dioxide is fed on a one-to-one ratio to chlorine residual, so the gas storage volume is based upon the average chlorine residual and the average flow. Gas feed systems are sized based upon the peak demand of sulfur dioxide. If a zero residual of chlorine is required, the sulfur dioxide feed system is controlled based upon the chlorine residual concentration prior to sulfur dioxide addition, with flow and residual pacing of the sulfur dioxide and sulfur dioxide feed rates calculated via a programmable logic controller. If a zero chlorine residual is not necessary, then a feedback loop control system is put in place, which paces sulfur dioxide based upon the residual concentration after sulfur dioxide addition and the flow rate of the water being treated with sulfur dioxide feed rates calculated via a PLC.

Carbon Dioxide Addition

Carbon dioxide can be used to lower the pH of water for the following process applications:

- Stabilization of lime softened water to eliminate excess precipitation of calcium carbonate and magnesium hydroxide, depending on the pH of the softening process
- Control of bromate formation during ozonation
- Optimize pH for coagulation
- Optimize pH for hexavalent chromium removal by weak base anion exchange
- Enhance the adsorption of arsenic (V) on iron media and iron floc
- Enhance the removal of sulfides in air stripping

Carbon dioxide and the carbonate system control the pH for most natural waters. The important carbonate species include carbonic acid (H_2CO_3), bicarbonate ($HCO^{-}{}_{3}$), and carbonate ($CO^{2-}{}_{3}$), the sum of which constitutes the total inorganic carbon of the water (CT). Bicarbonate and carbonate are the principal components of alkalinity in natural waters in the normal range of pH, while hydroxide ion (OH-) contributes to alkalinity in lime softening. Addition of carbon dioxide to water increases the CT and results in a decrease in the pH. Additional discussion of carbonate chemistry is provided in Chapter 4. The amount of carbon dioxide required to reduce the pH to a given value is influenced by the buffering capacity (i.e., alkalinity) and can be an expensive process for waters with high alkalinity. The addition of carbon dioxide does not change alkalinity or increase the sulfate or chloride content of the water as occurs with the addition of strong mineral acids.

For commercial use, carbon dioxide is marketed as either a gas or liquid under pressure, or as solid as dry ice. For use in water treatment processes, it is most commonly delivered to the facility by tanker truck and pumped into pressurized outdoor storage tanks. Figure 3-4 shows a photo of a carbon dioxide tank, including the controls. Evaporators are usually required to convert the pressurized liquid carbon dioxide to a gas that is dissolved in the process stream through diffusers.

Courtesy of Lowry Aeration Systems Inc.

Figure 3-4 A carbon dioxide storage tank and controls

Hazards associated with carbon dioxide use include asphyxiation, frostbite, and kidney damage or coma caused by pH shifts in the blood due to elevated or decreased levels of CO_2 in the bloodstream.

Mixing

Mixing with air is used at drinking water treatment plants primarily in situations where other mixing approaches (e.g., mechanical mixers) may be difficult to implement (e.g., channels between treatment process) or where air may be preferred (e.g., low DO water). When air mixing is used, it is important that the systems are designed to provide even air distribution. Applications involving diffused bubble aeration where other considerations such as efficient gas transfer kinetics are important are described in Chapter 2.

When air is injected or diffused into water, it imparts energy that provides mixing. The key parameter for mixing with air is the air flow rate per unit volume of the mixing chamber (e.g., scfm/1,000 cubic feet). Typically, the air is compressed in blowers and is then introduced into the mixing chamber at depth. When ambient air is used, it usually requires filtration.

Flocculation. Air mixing for flocculation is described in terms of velocity gradient G (sec^{-1}), which can be defined as follows:

$$G = (P/\mu)^{0.5} \quad \text{(Eq 3-1)}$$

where P is power dissipated by the rising air bubbles per unit volume and μ is the absolute viscosity of water. Fair, Geyer, and Okun (1968) provide the following equation to calculate P:

$$P = 81.5\ Q^{a} \log [(h+34)/34] \quad \text{(Eq 3-2)}$$

where Q^{a} is the air flow rate (standard cubic feet per minute) and h is the depth of the diffusers below the water surface. The actual G values used will depend on process requirements and goals.

Distribution system reservoirs. Bubble aeration can provide good mixing to eliminate dead spots or stratification in distribution system reservoirs if the diffuser system provides uniform coverage of the entire surface area. However, because many reservoirs are deep and water levels can change during the day, bubble aeration systems may be energy inefficient and hard to operate. In such cases, other aeration technologies including spray nozzles, jet aerators, and surface aerators may be more appropriate, especially if other treatment objectives, such as trihalomethane removal, are important (see discussion in Chapter 4). There are many commercial mixing products available to address specific applications. The degree of mixing achieved by aeration equipment requires site specific assessment. Computational fluid dynamics (CFD) and tracer studies (for existing facilities only) are two tools that may be useful in quantifying reservoir mixing when more detailed analysis is required for critical water system operation.

Solids equalization. Mixing of suspended solids with air can be a temperamental process. Typical range of air intensity for completely mixed flow regimes for wastewater aeration basins that keep solids suspended is 18–30 scfm/1,000 cubic feet (Metcalf and Eddy 1979, WEF and ASCE 2010). For suspensions with solids content greater than 10 percent, moderate to violent mixing, with air intensities of 72–340 scfm/1,000 cubic feet may be required for chamber depths of 10 ft (Perry and Green 1986), although more typical air intensities around 20 scfm/1,000 cubic feet have been satisfactory for deeper basins where additional energy is required to overcome the pressure due to greater depth.

REFERENCES

AWWA (American Water Works Association). 2010 (6th ed.). *Water Quality and Treatment*. New York: McGraw-Hill Book Co.

Fair, G.M.; Geyer, J.C.; and Okun, D.A. 1968. *Water and Wastewater Engineering. Volume 2*. Water Purification and Wastewater Treatment and Disposal. New York: John Wiley & Sons.

Kawamura, S. 1991. *Integrated Design of Water Treatment Facilities*. New York: John Wiley & Sons, Inc.

Lowry, J.D. and L. Saindon. 2013. Rapid Corrosion of a Stainless Steel Air Stripper by Trichloramine. presented at the 2013 California-Nevada Inorganic Contaminants Symposium, 2/5/-2/6/2013, Sacramento, CA.

Lytle, D.A.; White, C.; Williams, D.; Koch, L.; and Nauman, E. 2013. Innovative biological water treatment for the removal of ammonia. *Journal AWWA,* September, 105(9):87–88.

Metcalf & Eddy. 1979 (2nd ed.). Wastewater Engineering: Treatment Disposal Reuse. Revised by George Tchobanoglous. New York: McGraw-Hill.

Odell, L.H. 2010. Treatment Technology for Groundwater. Denver, CO: American Water Works Association.

Perry. R.H. and D.W. Green. 1986 (6th ed.). *Perry's Chemical Engineers' Handbook.* (Robert H. Perry, and Don W. Green, editors). New York: McGraw-Hill.

Reiber, S. 1993. Investigating the Effects of Chloramine on Elastomer Degradation. *Journal AWWA,* 85(8):101.

Rockaway et. al. 2007. Performance of Elastomeric Components in Contact with Potable Water. *Journal AWWA,* 99(12):99–110.

WEF and ASCE. 2010. Design of Municipal Wastewater Treatment Plants. WEF Manual of Practice No. 8 and ASCE Manuals and Reports on Engineering Practice No. 76. Fifth Edition. Alexandria, VA: WEF Press and New York: McGraw-Hill.

White, G.C. 1992 (3rd ed.). *Handbook of Chlorination and Alternative Disinfectants.* New York: Van Nostrand Reinhold.

This page intentionally blank.

Chapter 4

Gas Removal

The earliest use of gas transfer processes in drinking water was to remove undesirable dissolved gases. Engineering texts written early in the twentieth century advocate aeration of raw waters to "wash out gases and volatile odors" (Flinn et al. 1916). The importance of removing carbon dioxide and hydrogen sulfide was well understood at that time.

With the increasing use and distribution of gasoline, solvents, and other manmade organic chemicals after World War II, contamination of aquifers with synthetic organic chemicals became widespread. Many of these synthetic organic chemicals are volatile, so that gas transfer processes are used for the removal of dissolved gases both of natural and of anthropogenic origin.

APPLICATIONS

Modern gas transfer applications include removal of hydrogen sulfide, carbon dioxide, methane, radon, and regulated volatile organic chemicals (VOCs), primarily from groundwater. Some surface supplies also require air stripping to remove contaminants such as THMs and nitrogen trichloride.

There are many applications for aeration in drinking water treatment, such as for removal of naturally occurring constituents like hydrogen sulfide, radon, methane, and ammonia; manmade volatile organic compounds (VOCs); and disinfection byproducts (DBPs). Hydrogen sulfide is a common contaminant of groundwater in Florida, Texas, and Louisiana, and is found in wells throughout the United States (ASTDR 2006). Methane commonly co-occurs with hydrogen sulfide. United States Geological Survey sampling and analysis of 932 public supply wells revealed that more than 50 percent have radon at greater than 300 pCi/L (Toccalino and Hopple 2010). Ammonia contamination of groundwater has been reported at many locations around the country—from Cape Cod, Mass. (Böhlke et al. 2006) to Kansas (KDHE 2010) to California and elsewhere (Henrie et al. 2012).

The US Geological Survey (USGS) reported that 90 of 98 aquifer studies detected VOCs in the tested wells though only two percent of wells had one or more VOCs above the Safe Drinking Water Act Maximum Contaminant Level (Zogorski et al. 2006). Gas transfer is the best available technology for the removal of regulated volatile organic chemicals from water (USEPA 1987). More recently, water utilities have used aeration of surface and groundwaters in the distribution system to lower levels of DBPs.

REMOVAL OF AMMONIA

In water, inorganic ammonia nitrogen can be present as ammonium ion (NH_4^+) or ammonia (NH_3). These two species are present in equilibrium with ammonia (NH_3) at a pH of 9.25 (the acid dissociation constant, or pKa). Below pH 9.25, ammonium ion dominates, and above 9.25, ammonia is the primary species. The non-ionized (uncharged) ammonia species, not the ammonium ion, can be removed by aeration. The distribution of these species is given by the governing equilibrium equations shown below and illustrated in Figure 4-1.

Equation 1: Equilibrium between Ammonia and Ammonium Ion

$$NH_4^+ \leftarrow NH_{3(aq)} + H^+ \qquad pK_1 = 9.25 \text{ at } 25°C$$

Very few actual aeration systems are used for ammonia removal in drinking water because of the necessary increase in pH. Removal of ammonia by aeration typically requires raising water pH to between 10.8 and 11.5, the range in which non-ionized ammonia gas (NH_3, rather than ammonium, NH_4^+) is >97 percent of the total ammonia (USEPA 2000). Waters with even moderate carbonate hardness will scale when the pH is raised this high, similar to a lime softening process. Scaling can foul the aeration equipment, downstream pumps, and the piping in distribution systems. In aeration post-treatment, pH needs to be reduced to acceptable levels for distribution. Anti-scaling strategies may not be effective in controlling this problem.

The Henry's law constant for ammonia at 20°C is 0.76 standard atmospheres (atm), which is three orders of magnitude lower than the constant for hydrogen sulfide for example (Kavanaugh and Trussell 1980, Landolt-Boernstein 1976). Ammonia removal efficiency is highly dependent on air temperature and air-to-water ratios (USEPA 2000).

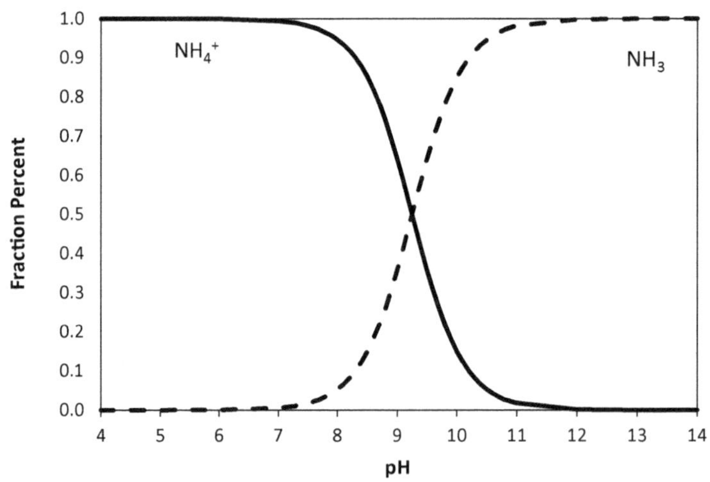

Source: Haynes 2014.

Figure 4-1 Inorganic ammonia nitrogen species in aqueous solutions with respect to pH

A high air-to-water ratio is required, compared to hydrogen sulfide for example. Other removal options for ammonia, such as breakpoint chlorination or biological treatment, are more commonly used for drinking water treatment (Blute et al. 2012).

REMOVAL OF CARBON DIOXIDE

Carbon dioxide (CO_2) is a colorless, odorless, incombustible gas that occurs in the atmosphere at approximately 0.027–0.039 percent. Carbon dioxide enters surface and groundwater through atmospheric absorption or through biological oxidation of organic matter. For the purpose of this manual, primary discussion of carbon dioxide will be based on its importance in drinking water aeration.

The solubility of CO_2 in water, according to Henry's Law, is between 0.3–0.8 mg/L when in equilibrium with the atmosphere. Concentrations of carbon dioxide in soil gas are approximately 100 times that in air, due to biological activity. Groundwater flowing through soil and rock formations dissolves carbon dioxide and is pumped to the surface at concentrations as high as several hundred mg/L. Typical concentrations in groundwater are between 10–100 mg/L.

A small fraction (<1 percent) of free carbon dioxide dissociates in water to form carbonic acid (H_2CO_3), which can dissociate further into H^+, HCO^-_3, and CO^{2-}_3 depending on pH. The governing equilibrium reactions for carbon dioxide, water, and the carbonate buffer system are given below and illustrated in Figure 4-2.

Equilibrium between dissolved CO_2 and H_2CO_3

$$\frac{[H_2CO_3]}{[CO_2][H_2O]} = 2.6 \times 10^{-3} \qquad \text{(Eq 4-1)}$$

Equilibrium between H_2CO_3 and HCO^-_3

$$\frac{[HCO^-{}_3][H^+]}{[H_2CO_3]} = 1.7 \times 10^{-4} \qquad \text{(Eq 4-2)}$$

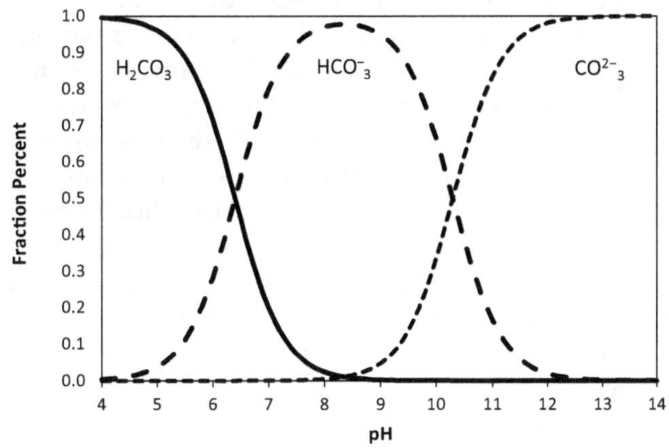

Source: Pedersen et al. 2014

Figure 4-2 Inorganic carbonate species in aqueous solutions with respect to pH

Equilibrium between HCO^{-}_{3} and CO^{2-}_{3}

$$\frac{\left[CO^{-}{}_{3}\right]\left[H^{+}\right]}{\left[HCO^{-}{}_{3}\right]} = 5.6 \times 10^{-11} \qquad \text{(Eq 4-3)}$$

Equilibrium of water

$$\frac{\left[OH^{-}\right]\left[H^{+}\right]}{\left[H_2O\right]} = 1.019 \times 10^{-14} \qquad \text{(Eq 4-4)}$$
$$\left[H_2O\right] = 1$$

Mass balance of carbon

$$\left[CO_2\right] + \left[H_2CO_3\right] + \left[CO^{2-}{}_{3}\right] + \left[HCO^{-}{}_{3}\right] = \left[DIC\right] \qquad \text{(Eq 4-5)}$$

Electroneutrality expression for a carbonate system

$$\left[HCO^{-}{}_{3}\right] + \left[2CO^{2-}{}_{3}\right] + \left[OH^{-}\right] = \left[H^{+}\right] + \left[Na^{+}\right] \qquad \text{(Eq 4-6)}$$

Total alkalinity of a carbonate system

$$\left[\left[HCO^{-}{}_{3}\right] + \left[2CO^{2-}{}_{3}\right] + \left[OH^{-}\right] - \left[H^{+}\right]\right] \times 50,000 = \text{Alk}\left(\frac{\text{mg}\,CaCO_3}{\text{L}}\right) \qquad \text{(Eq 4-7)}$$

CO_2 saturation

$$\left[CO_2\right]_{sat} = KH \times P_{CO_2} \quad KH = 3.889 \times 10^{-4}\,\frac{\text{mol}}{\text{kPa}} \qquad \text{(Eq 4-8)}$$

For a given temperature, the equilibrum equations can be expressed in terms of dissolved inorganic carbon (DIC), pH, and alkalinity. If two of these three parameters must be known, then the third can be solved. Typically, the alkalinity and pH are most easily measured, and the carbon dioxide is then calculated. Graphical and mathematical models are readily available for this purpose. DIC can also be measured but is the most inaccurate of the three parameters to determine. Care must also be taken in measuring the pH because any loss of carbon dioxide during sampling, transport or measurement will result in a lower than actual calculated concentration of total inorganic carbon. The alkalinity is the most easily measured parameter because it is unaffected by a loss of carbon dioxide or a change in pH. However, for waters with very low alkalinity (< 5–10 mg/L), careful titration of alkalinity with an accurate pH meter is required, rather than using an indicator solution for an endpoint determination. The most accurate total inorganic carbon determination is made by calculation from an accurately determined pH done in the field, and an alkalinity performed in the laboratory.

It is important to note that carbon dioxide dissolved in pure water can lower the pH to 5.65. As an example, a drinking water with an alkalinity of 100 mg/L and a carbon dioxide of 0.7 mg/L (atm equilibrium) will have a pH of approximately 8.3–8.4. Increased carbon dioxide will depress the pH to as low as 5.6, and it will take the carbon dioxide levels shown in Table 4-1 to do so. The carbon dioxide shown in Table 4-1 is available for removal by air stripping, down to the equilibrium level of 0.7 mg/L. This would return the water to a pH of 8.3–8.4. It should also be realized that aerating water with a pH above 8.3–8.4 will lower the pH through recarbonation. In summary, aeration can raise the pH by

Table 4-1 Amount of CO_2 required to depress the pH in one example

CO_2 (mg/L)	pH	Alkalinity (mg/L)
0.7	8.33	100
2.4	8.0	100
24	7.0	100
75	6.5	100
238	6.0	100
598	5.6	100

stripping excess carbon dioxide (with respect to air), and it can also be used to lower the pH if the pH is above 8.3–8.4.

CO_2 Removal for Corrosion Control

Aeration can be an effective method of corrosion control in water supplies containing significant CO_2 and low to moderate alkalinity. Both lead and copper may be reduced to lower levels by stripping CO_2 to raise the pH, without the use of any corrosion control chemical. Air stripping offers several advantages over adding alkalinity and/or phosphate for lead and copper control in distribution systems. These include more stability in the resulting treated water pH, no increase in alkalinity, no addition of the conjugate ion associated with the addition of a base, concurrent removal of other volatile constituents such as radon, and, in some cases, a lower operating cost.

For groundwater with significant CO_2, it is not uncommon for the pH to be depressed by 1.0–1.5 pH units, and up to 2.5 pH units in extreme cases. Depending on the pH and alkalinity, this means the water may contain more than 500 ppm of CO_2. A water with this type of chemical quality is not a good candidate for chemical addition to raise the pH. Neutralizing a high concentration of CO_2 with a base can result in excessive alkalinity and added Na^+ or K^+, and a net deterioration of water quality.

Carbon dioxide is removed in a manner similar to any volatile gas in water. Its ease of stripping is between radon and PCE, and it is considered moderately easy to remove using aeration. Removing carbon dioxide from groundwater raises the pH and reduces corrosivity. Oxygen can be corrosive in boiler applications, or in waters where the temperature is elevated. While oxygen is added in waters that are below saturation, the benefits of removing carbon dioxide are much more important in comparison to any small increase in corrosivity attributable to an increased oxygen level because carbon dioxide is far more corrosive during water distribution than oxygen at ambient temperatures.

Some water treatment processes also require the removal of carbon dioxide, and these include reverse osmosis plants and arsenic removal plants that use carbon dioxide to lower pH prior to membranes or granular filtration, and sulfide removal operations, where carbon dioxide may be used to enhance removal. In these operations, the air stripping of carbon dioxide is commonly referred to as decarbonation.

Although carbon dioxide removal via aeration has been used for centuries, increased application of the process occurred as a result of the Lead & Copper Rule. There are many groundwater systems that have excess carbon dioxide and a depressed pH. Initially, chemical solutions were used, but it is generally true that air stripping is a superior treatment because it often produces a superior finished water quality in terms of alkalinity and stability and maintenance of the final treated water pH. In many cases, air stripping has proven to be more cost-effective and reliable.

Schock et al. (2002) described a representative case study of a multi-staged bubble air stripper installed at two utilities in CA. The utilities had been unsuccessful over a 4-yr period using chemical addition to achieve corrosion control. Following implementation of air stripping, the pH was raised from 6.1–6.4 to 7.5–7.8 (96 percent CO_2 removal), the dissolved inorganic carbon was reduced by half, and lead and copper were reduced to well below Action Levels in less than 4 months. Further, there were no detrimental effects of increased dissolved oxygen.

Lowry (2010) reported on over 20 years of experience in air stripping carbon dioxide for corrosion control, using the multi-staged bubble system described in Chapter 2. Water supplies containing as low as 8 mg/L and as high as 200+ mg/L alkalinity have been successfully treated without the addition of corrosion control chemicals. The conclusions of that work were

- Air stripping can be an ideal solution to corrosion control problems caused by high carbon dioxide in groundwaters with low to moderately high alkalinity
- Corrosion control chemicals can be eliminated or avoided by using air stripping, resulting in a superior water quality following treatment
- Air stripping CO_2 to low levels produces a treated water with a more stable pH and water quality, as compared to a chemically-treated water
- Normally, the pH needs to be raised to above 7.2 to achieve low copper solubility, and above 7.5–8 for reducing lead in copper pipes with lead solder or fixtures

It should be noted that it normally takes much longer to achieve reduction in lead levels, as compared to copper, following the implementation of air stripping for corrosion control. It can take as little as a few hours to eliminate copper violations in distribution pipes, and as long as several months to do the same for lead.

CO_2 Removal for Post-Treatment

Some water treatment processes also require the removal or addition of carbon dioxide, including reverse osmosis plants (a process called decarbonation), arsenic removal plants that use carbon dioxide to lower pH prior to filtration, and sulfide removal operations, where carbon dioxide is added to enhance sulfide removal.

Removal of Sulfides

Inorganic sulfides in drinking water consist of hydrogen sulfide (H_2S), bisulfide ion (HS^-), and sulfide ion (S^{2-}), all in equilibrium as a function of the pH. The pKa for the distribution of H_2S/HS^- is 7.00 at 25°C. The pKa for HS^-/S^{2-} is 12.92 at 25°C. This means that the important sulfide species present in some groundwaters are HS^- and H_2S, not S^{2-}. The distribution of these species is given by the governing equilibrium equations shown below and illustrated in Figure 4-3.

$$H_2S + H_2O \Leftrightarrow H_3O^+ + HS^-_{(aq)} \qquad pK_1 = 7 \tag{Eq 4-9}$$

$$HS^- + H_2O \Leftrightarrow H_3O^+ + S^{2-}_{(aq)} \qquad pK_2 =\sim 12-13.8 \tag{Eq 4-10}$$

Sulfides are generated and exist in reduced environments, such as aquifers with anoxic conditions and lake or ocean sediments. Sulfur-reducing bacteria can use sulfate as an electron acceptor and produce sulfides. Dissolved iron and manganese are often in waters containing sulfides. Sulfides are not normally associated with surface waters that are more aerobic. As a result, most drinking water supplies that contain objectionable sulfides are from well sources.

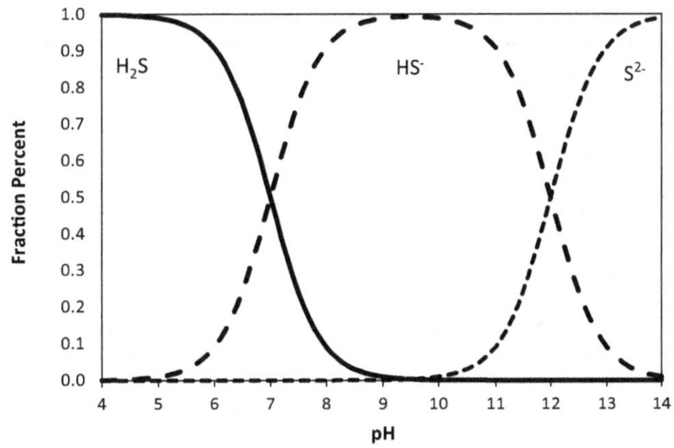

Source: Siekkinen et al. 2006

Figure 4-3 Sulfide species in aqueous solutions with respect to pH

Hydrogen sulfide is a colorless, flammable, and dangerous gas. It causes objectionable taste and odor issues (Overbeck 1995). It has an unpleasant, sulfurous taste and "rotten egg" odor with an odor threshold of approximately 0.001 mg/L (Suffet 1983). In water, H_2S can increase corrosion rates at mg/L levels (Hawthorn 1970, Jacobs et al. 1998) as well as increase turbidity, and cause maintenance issues throughout distribution systems and at water treatment plants (Duranceau et al. 2010). In air, the rate of corrosion is much higher, and it rapidly corrodes unprotected steels and stainless steels. These are the reasons for removing H_2S from drinking water, even though there is no primary MCL. All sulfides (H_2S and HS^-) should be completely removed so that reformation of hydrogen sulfide gas does not occur (Lyn and Taylor 1992).

Hydrogen sulfide is a deadly gas at elevated concentrations in air. A few breaths at 700 mg/m^3 can be fatal (WHO 2003). These levels are much higher than those typically associated with drinking water supplies and operations; however, safety should always be taken into consideration when removing H_2S from water. Accidental deaths from inhalation of H_2S occur in municipal and industrial wastewater operations and in oil field operations. The addition of cleaning acid to tanks containing sulfides can suddenly release very high concentrations of hydrogen sulfide gas into the air.

Methods of Hydrogen Sulfide Removal

Field experience from numerous sources indicates that the H_2S fraction is easily air-stripped and that the remaining HS^- does not convert to H_2S during aeration, due to a rise in the pH during stripping caused by the concurrent removal of carbon dioxide. Typically, the natural alkalinity present results in a pH in the range of 8.0–8.4 after the removal of carbon dioxide. Selection of aeration methods depends on the total sulfide concentrations and initial pH, as summarized by the Florida Blackwater Rule, Florida Administrative Code, Chapter 62-555, shown in Table 4-2.

The Florida Blackwater Rule and many references in the literature indicate that the pH must be lowered with acid to effect a greater removal of sulfides. As summarized in the Blackwater Rule, acid addition is a common method to increase the rate of sulfide removal. This allows the aeration equipment to remove more sulfides and greater than 99 percent removal is possible. The level of sulfide in water does not influence the percent efficiency of a given type of aeration equipment, as the rate of reaction is always first order. However, a higher initial sulfide concentration will result in a higher final concentration for the same removal percent, which may be objectionable to customers. With acid providing the

Table 4-2 Florida blackwater rule (State of Florida)

Potential for Impacts without Total Sulfide Removal	Water Quality Ranges	Potential Water Treatment
Low	Total sulfide < 0.3 mg/L Dissolved iron < 0.1 mg/L*	Direct chlorination†
Moderate	0.3 mg/L ≤ Total sulfide ≤ 0.6 mg/L @pH ≤ 7.2	Conventional aeration‡ (maximum removal efficiency ≈ 40–50%)
	Or	Or
	0.3 mg/L ≤ Total sulfide ≤ 0.6 mg/L @pH > 7.2	Conventional aeration with pH adjustment§,** (maximum removal efficiency ≈ 40–50%)
Significant	0.6 mg/L < Total sulfide ≤ 3.0 mg/L @pH ≤ 7.2	Forced draft aeration‡ (maximum removal efficiency)
	Or	Or
	0.6 mg/L Total sulfide ≤ 3.0 mg/L @pH > 7.2	Forced draft aeration with pH adjustment§,** (maximum removal efficiency ≈ 90%)
Very significant	Total sulfide > 3.0 mg/L	Packed tower aeration with pH adjustment§,** (maximum removal efficiency > 90%)

* High iron content raises concern if chlorination alone is used and significant dissolved oxygen exists in the source water. Filtration may be required to remove particulate iron prior to water distribution.
† Direct chlorination of sulfide in water in the pH range normally found in potable sources produces elemental sulfur and increased turbidity. Finished-water turbidity should not be more than two nephelometric turbidity units greater than raw-water turbidity.
‡ Increased dissolved oxygen entrained during aeration may increase corrosivity.
§ Reduction of alkalinity during pH adjustment and high dissolved oxygen entrained during aeration may increase corrosivity. Corrosion control treatment such as pH adjustment, alkalinity recovery, or use of inhibitors may be required.
** High alkalinity will make pH adjustment more costly and use of other treatment may be in order. Treatment that preserves the natural alkalinity of the source water may enhance the stability of finished water.

protons to convert HS^- to H_2S, any level of treatment can be achieved with proper sizing of the treatment process.

Mineral acids, such as sulfuric acid (H_2SO_4) and hydrochloric acid (HCl), and carbon dioxide (CO_2) are often used as pretreatment for aeration to supply the protons needed for converting HS^- to H_2S. Sulfuric acid is often utilized for relatively low cost, effectiveness, and availability; however, it is corrosive and demands special handling. Strong acids also reduce alkalinity, which can either be a positive (reduced scaling) or a negative (increased corrosivity for certain water qualities). Post-treatment may be required to stabilize water. Alternatively, carbon dioxide can be used for pH reduction without decreasing alkalinity.

The Henry's law constant for hydrogen sulfide at 20°C is 5.15×10^2 standard atmospheres (atm) (Kavanaugh and Trussell 1980, Landolt-Boernstein 1976). This is very close to the Henry's constant for trichloroethylene (TCE). For context, benzene has a slightly lower Henry's constant of 3.53×10^2 atm, and oxygen has a Henry's constant of 4.3×10^4 atm (Kavanaugh and Trussell 1980, Landolt-Boernstein 1976).

Packed tower aeration can be effective for water with high sulfide levels depending on the pH, with removal efficiencies of up to 99 percent (Duranceau 2010, Kavanaugh and Trussell 1980). Similar to forced draft aeration, packed tower aeration can be susceptible to microbiological growth in the packing material. Deposition of calcium carbonate on the packing material can also occur in hard waters, requiring either cleaning or use of a sequestering chemical.

Lowry (2010) reported on a multi-staged bubble aeration system that was designed to treat a 3–12 mgd groundwater source in Florida. This work was preceded by a 550-gpm pilot study at another site in Florida, where it was observed that CO_2 added to the first few stages of treatment significantly lowered the required air-to-water ratio (A:W ≈2.0) to well below what would be expected for a high removal of sulfides based on the Henry's constant for H_2S. The larger plant was designed accordingly and greater than 99.9 percent sulfides (12,000 ppb to 10 ppb) were removed with CO_2 added to stages 1 and 2 of the six stages in the basin. Carbon dioxide was added to the bubbles, rather than to the inlet water. In this way, the first stages removed sulfides without raising the pH, and the remaining stages raised the pH by stripping the excess carbon dioxide. The minimum pH in the first stages was slightly above 7, demonstrating that complete removal of sulfides at a very high rate is possible without significantly lowering the pH according to equilibrium. The same removal could have been achieved for a higher pH source water, as long as the same amount of carbon dioxide was used in additional stages. In the design of multi-staged bubble systems, both the air intensity (AI) described in Chapter 2, and the CO_2 intensity (CDI) are important for the effective removal of sulfides.

Off-Gas Treatment From Sulfide Removal

The United States Environmental Protection Agency (USEPA) currently has no off-gas standards for hydrogen sulfide. Individual states have varying levels of guidance with regard to hydrogen sulfide for air quality control. For example, the Florida Department of Environmental Protection (FDEP) references guidelines for objectionable odors.

The Occupational Safety and Health Administration (OSHA) provides recommendation for a threshold of 20 parts per million by volume (ppmv) for on-site hydrogen sulfide off-gas concentrations. Other sources recommend a threshold of 7.3 ppmv at the property boundary for avoiding odor complaints (Duranceau 2010).

California has an ambient air quality standard for hydrogen sulfide. The California Board of Public Health adopted an air quality standard in 1962, where a hydrogen sulfide level of 0.1 ppm for one hour was deemed an "adverse" level. Since then, the Air Resources Board adopted a standard of 0.03 ppm of hydrogen sulfide for a one-hour average in 1969, and this standard was retained in 1984 as well (California EPA Air Resources Board 2009).

Scrubber systems are often used to remediate the odor generated from aeration treatment technologies, such as packed tower aeration and forced draft aeration. Typically, granular activated carbon (GAC) scrubbers are used for removing hydrogen sulfide from the air. GAC scrubbers are effective at low concentrations, and for higher concentrations of about 5 to 25 ppmv, chemical scrubbing in packed towers is effective. Selection of scrubber systems should be based on hydrogen sulfide concentration and corresponding operations and maintenance costs. The scrubber systems can be costly, thus having an impact on the overall treatment plant operations.

REMOVAL OF RADON

Radon is an odorless, colorless noble gas. It is found naturally in soil and rock formations and, due to volatility, can be transferred into the atmosphere, buildings, and groundwater supplies. Radon is one of the contaminants that EPA regulates under the Safe Drinking Water Act. Groundwater radon concentrations range from less than 100 pCi/L to greater than 1,000,000 pCi/L and depending on geology, with high concentrations often associated with granitic formations. Most groundwater supplies contain less than 1,000 pCi/L.

Radon is produced through the decay of radium, through the uranium and thorium decay chains. Table 4-3 summarizes the general physical and chemical properties of ^{222}Rn. This isotope has a number of unique properties that utilities should consider when selecting and applying radon removal technologies, including:

- It does not interact with other elements or compounds except in unusual circumstances that do not occur in drinking water supplies.
- It occurs at very low mass concentrations (e.g., 1,000 pCi/L = 6.5×10^{-10} μg/L).

Table 4-3 General physical and chemical properties of radon

Property	Value
Molecular weight	222 g/mol
Radioactive half-life	3.82 days
Specific activity	1.54×10^5 pCi/pg
Boiling point	211 K (-80°F)
Melting point	202 K (-96°F)
Solubility in water	51 cm^3/L at 0°C (32°F)
	8.5 cm^3/L at 60°C (140°F)
Molar volume	0.050 m^3/kmol
Air diffusion coefficient (20°C)	2.0×10^{-5} m^2/s
Water diffusion coefficient (20°C)	1.4×10^{-9} m^2/s
Henry's law constant (20°C)	4.08 (dimensionless)

Sources: Cummins 1988, McKone 1987, and USDHEW 1970 (Table from Drago 1998)

- It is relatively volatile, with a Henry's law constant (4.08 at 20°C) between oxygen (30 at 20°C) and carbon dioxide (0.9 at 20°C).
- It is the decay product of radium-226 (^{226}Ra) and is an alpha particle emitter with a half-life of 3.82 days.
- The primary health risk is from inhalation of ^{222}Rn and its short-term decay products (half-lives all less than 30 minutes) transferred from water to air (10,000 pCi/L of water is equivalent to 1 pCi/L of air).

^{222}Rn enters groundwater as a result of the radioactive decay of members of the uranium series (starts with ^{238}U and includes ^{234}U and ^{226}Ra) in aquifer materials. Some of the ^{226}Ra and/or ^{238}U and ^{234}U may dissolve into the groundwater and co-occur with ^{222}Rn.

Radon is principally a groundwater concern because it is typically aerated from surface water sources by natural processes. In equilibrium with the atmosphere, radon in water will be less than 1 pCi/L.

Technologies for radon removal are currently limited to aeration and liquid phase granular activated carbon (GAC). However, because radon is very volatile and readily transfers from water to air, the EPA has identified aeration technologies as "best available technology" (BAT) for radon removal.

Methods of Radon Removal

A number of existing and emerging aeration technologies are available for removing radon from groundwater. Much of the early research on removal of radon from groundwater by aeration was conducted at the University of Maine (Lowry 1988, Lowry and Lowry 1988, Lowry and Brandow 1985, Lowry et al. 1984) and the University of New Hampshire (Kinner et al. 1989, 1990). USEPA has identified packed tower aeration (PTA), multi-staged bubble air stripping (MSBAS), slat tray aeration, and spray aeration as possible best available technologies.

Other emerging technologies that may be suitable for radon removal include jet aeration, mechanical surface aeration, crosscurrent PTA, sieve tray aeration, cascade step aeration, pressure aeration-degasification, and microporous hollow fiber membranes. In addition, simple, low technology techniques such as blending, losses in distribution systems prior to the first customer (if allowed), naturally ventilated (atmospheric storage), and passive losses during simple aeration (e.g., over weirs) in storage reservoirs and standpipes may be appropriate approaches, especially for very small systems, when only modest radon removal goals (20 to 50 percent) are required.

PTA and MSBAS technologies have been most frequently used to remove radon and radon removals of greater than 99.9 percent have been documented for these systems. Radon removal effectiveness of PTA and MSBAS technologies are well known, and in most cases, pilot testing would not be warranted. Information on radon removal by other technologies is included in *Critical Assessment of Radon Removal Systems for Drinking Water Supplies* (Drago 1998).

PTA. As discussed in Chapter 2, the important design criteria for PTA systems include the air-to-water ratio, hydraulic loading rate, and the packing height. Modeling of PTA systems, described in Chapter 2, requires determination of an overall mass transfer rate (K_La) either by experiment or estimated by theoretical or empirical correlation. In general, K_La increases as the hydraulic surface loading rate (v) increases.

Estimates of K_La using data on PTA facility performance from Kinner et al. (1990), Deb (1992), Cummings (1988), and Dixon et al. (1991) range from 22 to 142 h^{-1} and follow the relationship below:

$$K_La = 4.58v^{0.76}$$

where K_La is in units of h^{-1} and v is in units of gpm/ft^2 (Drago 1998). This dataset covers a wide range of influent radon concentrations (1,790–226,400 pCi/L), air-to-water ratios (2:1 to 84:1), hydraulic loading rates (8.5–50 gpm/ft^2), packing heights (10–17 ft), and radon removals (93.2–99.1 percent). Hiltebrand, Dyksen, and Raman (1987), citing data from Lowry, reported that, for a PTA unit operated at 30 gpm/ft^2 and an air-to-water ratio of 15:1, radon removal increased from 50 to 99.9 percent as packing height increased from 3.5 ft–33 ft.

MSBAS. The MSBAS process is a modified diffused aeration system where the water depth is shallow (sidewater depths from 1.5 to 3 ft as compared with 10–20 foot depths in typical aeration basins) and uses several compartments in series (see Chapter 2 for details). The process can be modeled as completely mixed stages in series (see Chapter 2) and removal rate is first-order. Detention time is generally not a design criterion as long as a desired air-to-water ratio is achieved.

Estimates of the first-order rate constant (k) using MSBAS performance data from Lowry et al. (1984), Perkins and Hewitt (1988), Kinner et al. (1990, 1993), and Dixon et al. (1991) range from 1.18 to 281 h^{-1} and follow the relationship below:

$$k = 36.9\ \phi^{0.80}$$

where k is in units of h^{-1} and ϕ is volumetric air intensity in units of cfm/ft^3 (Drago 1998). This dataset covers a wide range of influent radon concentrations (1,658–79,999 pCi/L), air-to-water ratio (0.7:1 to 120:1), volumetric air intensities (0.012–11.4 cfm/ft^3), water depths (1–6.3 ft), number of compartments in series (one [batch test] to six), detention times (1.46–60.3 min), and radon removals (69.3–99.4^+ percent). Some of the highest radon removals were achieved at low water depths (1.3–1.5 ft) with either relatively high air-to-water ratio (22.1:1) and volumetric air intensity (7.61 cfm/ft^3), and short detention time (2.43 min) or with very high air-to-water ratio (120:1), low volumetric air intensity (0.166 cfm/ft^3), and moderately long detention time (13.3 min). Long detention times do not necessarily guarantee high percent radon removal. For example, Rafferty (1963) reported 97 percent radon removal for a single (plug flow) basin with a depth of 4 ft, an air-to-water ratio of 8:1, a low volumetric air intensity (0.166 cfm/ft^3), and long detention time (24 min). In this case, the system appears to have achieved equilibrium.

Off-gas treatment. A major technical siting consideration for aeration facilities involves dealing with state and local air emissions regulations. This consideration is particularly important because there currently is no proven technology, including vapor phase GAC, for removing radon from offgas emitted by aeration systems. However, the national average background level for radon in air is 0.4 pCi/L. Using the generally acceptable conversion of 10,000:1 for radon in water to air, this is equivalent to water with 4,000 pCi/L. Thus, the contribution from most aeration facilities to the surrounding air would be below background, and no significant increase in outdoor air would be measurable. Nevertheless, in Nevada, California, New Jersey, and Pennsylvania, some risk assessments for drinking water aeration facilities have also included radon that was emitted concurrently with regulated VOCs. Thus, risk assessments may play a major role for some sites.

Narasimhan et al. (2005) conducted a study that included extensive measurement of radon and radiation dosimetry at 18 radionuclide (U, Ra, and Rn) removal plants in the United States. The radon levels from air vents were found to be rapidly diluted by ambient mixing. Radon levels were near background (≈0.5–1.5 pCi/L background) at distances of 6 ft. It was concluded, based on these data and modeling, that exposure to Rn levels at treatment plants is not a problem, with prudent location of air vent. Air vents must be located to avoid the air blower intake and any outdoor walls or barriers that would trap the discharged air in a small area.

REMOVAL OF TRIHALOMETHANES

Aeration has long been recognized as an effective technology for removing the more volatile trihalomethanes from drinking water. Aeration literature and research on TTHM reduction became prevalent following establishment of the 1979 TTHM Rule and subsequent Stage 1 DBP Rule. The Stage 2 DBP Rule with compliance requirements of maintaining TTHM concentrations below 80 μg/L on a locational running annual average (LRAA) basis has brought aeration back into consideration as a method to remove TTHMs and achieve distribution system compliance with the new regulation.

The TTHM requirements of the Stage 2 DBP Rule have been especially problematic where elevated summer temperatures drive the TTHM formation reaction between chlorine disinfectant and total organic carbon (TOC). Systems can be particularly challenged by TTHM requirements especially on the outskirts of the distribution system or consecutive systems where there is long water age, extended contact time with chlorine, and low flow rates through infrastructure sized for future development that has been delayed.

For many systems, the water treatment plant finished water reservoir and downstream distribution system reservoirs are the only locations available to address TTHM control with aeration. Installing enhanced coagulation, activated carbon, alternative disinfection, or other TOC reduction measures have high capital and annual operations costs. In contrast, aeration has demonstrated to have lower capital and annual operations costs and can be a very effective strategy for TTHM reduction. However, all aeration systems are not equally effective for TTHM reduction. Table 4-4 presents aeration technologies and a description of characteristics for TTHM reduction.

The recommended aeration approach for a system is dependent upon the source water quality, infrastructure type, and operations characteristics. The references included for this section present a list of completed and ongoing efforts which have provided important answers related to implementation of aeration strategies for TTHM control and also identified considerations that need to be further evaluated to help utilities best achieve compliance. This section discusses these items further.

- Tools for Sizing Aeration Processes: Tools have been developed to size aeration systems and predict the reduction of contaminants including TTHMs as listed in Table 4-5.

Table 4-4 Aeration technologies and removal of TTHMs

Aeration / Technology	Description
Spray aeration with ventilation	• Reservoir flow passed through spray nozzles • Active air ventilation recommended as part of strategy • Calculation tools available to size nozzles and flow rate for TTHM reduction • Energy consumption for pumping through spray system becomes important component of annual cost for large reservoirs
Surface aeration with ventilation	• Floating aeration units supported within reservoir • Active air ventilation recommended as part of strategy • Calculation tools available to size aerator horsepower for TTHM reduction • Technology was identified as a very energy-efficient process for TTHM reduction.
Bubble	• Blower used to pass air through diffuser system at bottom of reservoir • Calculation tools available to size blower for TTHM reduction • Can require significant energy to apply bubbles at bottom of tall water column which is not an energy-efficient process for TTHM reduction.
Tray/Packed tower	• Blower used to pass air countercurrent to water flow through tower system • Equipment must be installed external to the reservoir and requires pumping back to water storage
Active ventilation	• Supply fan passes filtered air into reservoir which is exhausted through existing vents. • Remove TTHMs from the headspace above water surface within the clearwell or reservoir • This low-cost strategy can be particularly effective when implemented on well-mixed reservoirs
Mixing	• Many reservoir mixing systems are available • Mixing systems help to remove dead zones within a reservoir where extended water age leads to increased TTHM formation

Table 4-5 Aeration strategies for reduction of TTHMs

TTHM Aeration Strategy	Analysis Tool
Spray aeration	Tool developed as part of Water Research Foundation Project 4413. Additional basis provided in AWWA Water Quality and Treatment Handbook
Surface, bubble, and packed tower aeration	Aeration System Analysis Program (ASAP) developed by Michigan Tech

- Impact on Chlorine Residual: Hypochlorous acid (HOCl) and hypochlorite ion (OCl^-) are the predominant chlorine species present in water with pH in the range between 6.5 and 8.5. These compounds have extremely unfavorable Henry's constants. HOCl is less strippable than ethanol and both are considered to be non-strippable from drinking water. Full-scale testing has shown very minor loss of chlorine residual with operation of aeration systems.
- Increasing pH: Data from full-scale aeration installations show a small increase of approximately 0.3 pH units with implementation of aeration processes for TTHM reduction. This pH change can lead to scaling formation on aeration equipment. Maintenance to remove scaling must be considered for implementation of aeration

strategies, if water quality modeling indicates a scaling potential. This problem is much more common in higher carbonate hardness groundwater and is not encountered as frequently with most surface water supplies.

- Reformation: Since Stage 2 DBPR compliance is based on concentrations at downstream distribution system sites, TTHM reformation is an important consideration. Data was collected from prior aeration installations which indicate that the concentration of TTHMs removed by aeration is approximately maintained over the remainder of the formation potential curve (see Figure 4-3).
- Reduction of Chlorinated and Brominated TTHMs: The chlorinated TTHMs (chloroform and dichlorobromomethane) are more volatile than the brominated TTHMs (bromoform and dibromochloromethane). Toxicology studies have indicated the brominated TTHMs have higher cancer risk, and as result, their removal is important. As shown in Figure 4-4, data collected from prior aeration demonstration testing confirms that brominated TTHMs (dibromochloromethane) are also removed by aeration.
- Optimized location for aeration systems: The location for aeration systems must be optimized within a utility distribution system because aeration can only remove TTHMs after formation. Typically aeration systems are located where TTHM concentrations are approaching the maximum concentration level (MCL) and there is short downstream water age.
- Active air ventilation: Data from full-scale tests has shown that when active ventilation systems are turned off or air flow rate is too low the result is poor performance of aeration systems for TTHM reduction. Some utilities have successfully achieved TTHM compliance by only adding active air ventilation to a well-mixed reservoir. This can be a low-cost strategy that should be considered. Further study on this topic is needed to develop predictive tools for sizing and

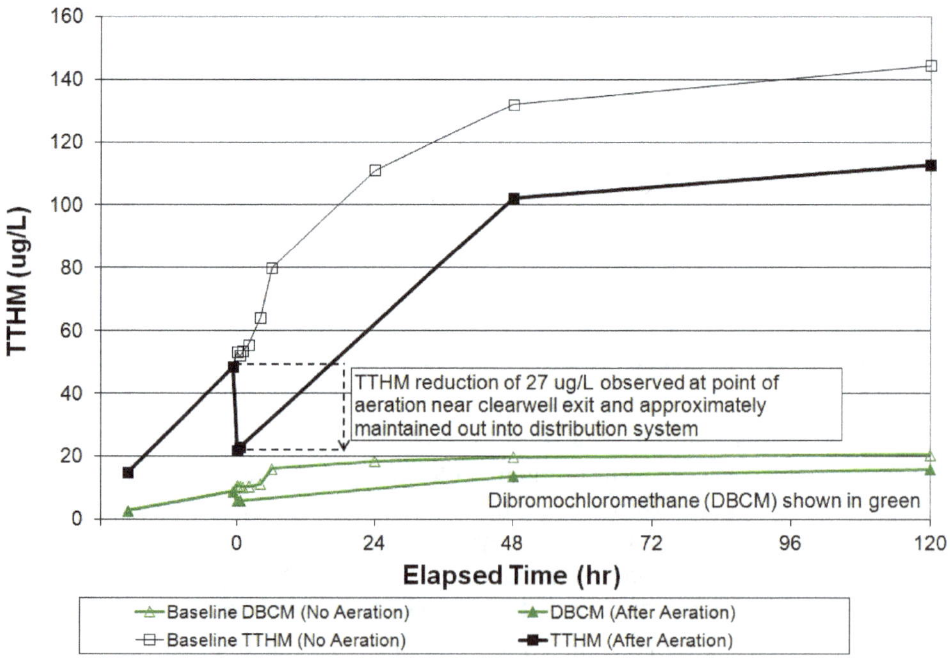

Note: Data obtained for City of Tempe, Ariz., aeration evaluation. Elapsed time of zero hours represents time when water leaves the finished water reservoir and enters the distribution system.

Figure 4-4 Aeration and TTHM reformation evaluation

performance of this strategy. The air flow rate achieved by active ventilation is dependent upon the degree of plugging of intake air filters which must be maintained for optimal system performance.

The following case studies from Phoenix, AZ and Southwest Harbor, ME are included in this section as examples of how utilities are using aeration technologies to help achieve Stage 2 DBPR compliance.

CITY OF PHOENIX, ARIZ., CASE STUDY

A distribution system optimization program conducted by the City of Phoenix investigated aeration alternatives for TTHM control at reservoir sites. Four aeration strategies were evaluated including three in-reservoir strategies (spray, bubble, and surface aeration) and one external strategy (tray tower aeration). The evaluation considered the following criteria: TTHM reduction, lifecycle cost, constructability, ease of operation, required time out of service, and mixing.

A summary of evaluation parameters is provided in Table 4-6.

The Aeration System Analysis Program developed by Michigan Tech was used to size bubble and surface systems for the target percent TTHM reduction. A similar sizing analysis for spray aeration was conducted. Equipment supplier estimates were utilized for the tray tower aeration system. The following is a summary of the evaluation results.

TTHM reduction: All aeration technologies could meet the 30% TTHM reduction target.

Cost: All aeration technologies can be constructed at relatively low capital cost compared with other TTHM reduction technologies. Annual operations cost for energy is a differentiator because surface aeration can achieve similar TTHM reduction with a smaller motor.

Constructability: In-reservoir strategies required that the reservoir be taken out of service and drained for equipment installation. This was acceptable for this system. External strategies require additional footprint which was available at the reservoir site.

Ease of operation: All systems require maintenance for motors. Aeration systems are susceptible to maintenance effort for scale removal.

Mixing: The evaluated in-reservoir aeration strategies will improve mixing for the reservoir.

The evaluation led to the selection of surface aeration based on relatively low energy use and lifecycle cost. The operations staff also preferred surface aeration due to the ease of operation, and the capability to remove and reinstall units without taking reservoirs out of service. Equipment installation was coordinated with scheduled tank rehabilitation efforts to minimize the capital cost. A picture of the equipment installation which consists of a single 7.5-hp aerator is shown in Figure 4-5.

The following observations have been obtained over multiple years of operation of this full-scale installation.

Table 4-6 Summary of Phoenix, Ariz., optimization program evaluation parameters

Parameter	Tray	Bubble	Spray	Surface
Reservoir volume, MG	2			
Flow rate, mgd	0.7			
Average water level, ft	25			
Target % TTHM reduction	30%			
Estimated Energy Use (kWh/yr)	210K	196K	299K	49K

Source: City of Phoenix Water Services Department

Source: City of Phoenix, Ariz.

Figure 4-5 Reservoir surface aerator installation

- The distribution system sample point with less than four hours water age in the downstream pressure zone from the reservoir averages 23 percent reduction of TTHMs between periods when the aeration system is shut down and brought online. This TTHM reduction has been measured over periods when TOC and chlorine residual values remained nearly constant.
- The reservoir provides storage on a distribution system pressure zone with a pipe that has bidirectional flow based on demand cycles. As a result, it is difficult to measure inflow and outflow TTHMs. The benefit of aeration on TTHMs is most clearly tracked when monitoring downstream distribution system sampling sites for periods when the aerator is switched from off to on operation.
- Residence time within the reservoir can fluctuate between about 5 and 25 hours depending on the level set point and water demand. Longer reservoir residence time allows for greater contact with the aerator which increases TTHM reduction. Before implementation of aeration the operations staff focused on reducing water age (typically by reducing reservoir level) for TTHM control. Implementation of aeration allows the operations staff greater flexibility to use full capacity of the reservoir with less impact for TTHMs in the downstream distribution system.
- Reservoir headspace ventilation is an important part of the TTHM reduction strategy. The system includes a filtered air supply with 1 hp fan. Windblown dust events can clog the air filter which reduces air flow through the reservoir and decreases efficiency of the aeration system for TTHM reduction. Maintenance of the air filters is essential for optimal performance of the system.

TOWN OF SOUTHWEST HARBOR, MAINE, CASE STUDY

The Town of Southwest Harbor, Maine, uses a pristine lake as the source of its 1 MGD water supply. Chlorine is used as the disinfectant and the Town struggled for several years with MCL violations for both TTHMs and HAA5s. An engineering team was hired to evaluate possible solutions, with the goal of avoiding a conversion to chloramination. Oliver Associates and Lowry Engineering did the evaluation and aeration was the most attractive candidate for providing a solution.

The existing water system is unique because there is a coagulation and direct filtration plant and the downstream chlorine contact basin is greatly oversized for future flow rates that have not materialized. The chlorine contact basin is a nine (9) pass serpentine design and 15–20 ft deep. Existing TTHM data indicated that significant formation occurred in the chlorine contact basin because it was oversized. With further evaluation and testing to gather data for the design of a multi-staged bubble aeration process, it was decided that treatment could be effective, if the ninth pass of the contact basin was converted to aeration. The water depth in the last pass was reduced from 15 to 4 ft so that the multi-staged process could be more efficient. The required depth for chloroform removal is relatively shallow and depths in excess for 3–4 ft are extremely inefficient because compressors, rather than centrifugal blowers, would make the cost of providing the air prohibitive. The ninth pass was fitted with stage dividers and overflow weir and the system was commissioned. The system achieves >95 percent removal of TTHMs and has provided a solution to the Town's MCL violations. There is additional formation in distribution, especially in the winter when demand is low, but the system is meeting the TTHM MCLs. HAA5s were also significantly lowered with a pH change in the operation of the upstream filtration process.

Details of a stage are shown in Figure 4-6. Aerators are at 3.5 ft submergence and cover the entire floor of the aeration channel. There are nine (9) completely mixed stages in the 45-ft long channel.

An interesting observation was made after operating the aeration system for several weeks. The operator reported a significant loss of chlorine residual in the water exiting the chlorine contact basin. Since chlorine is not air-stripped, unless there is trichloramine present, it has been concluded that the intense mixing in the channel has resulted in more chlorine utilized in the formation of TTHMs. This is a benefit because the goal is to have a maximum formation of TTHM prior to aeration.

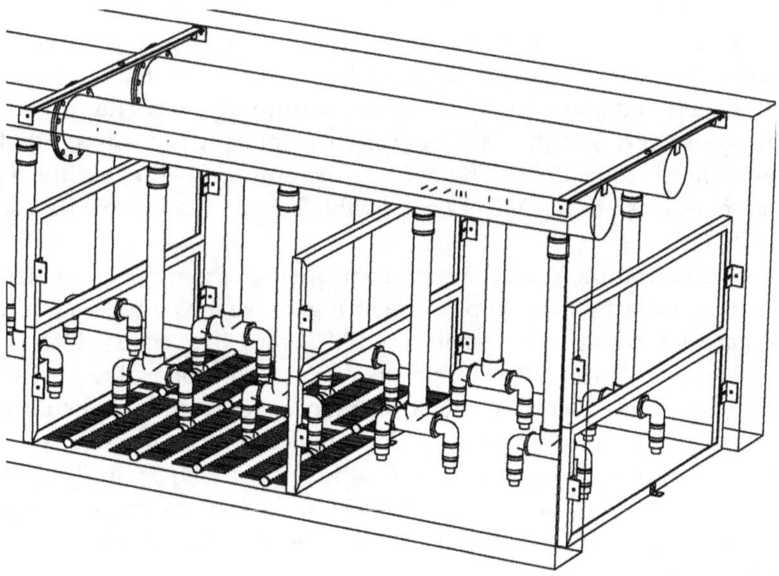

Source: Lowry Aeration Systems, Inc.

Figure 4-6 Details of a stage in the 9-stage channel system

REMOVAL OF VOCs

Occurrence of VOCs in Drinking Water Supplies

Volatile organic compounds (VOCs) in drinking water supplies can be attributed to a variety of anthropogenic sources including but not limited to hazardous waste sites, landfills, leaking underground storage tanks, underground injection wells, and agricultural activities (Ram 1990). Some hydrocarbon VOCs can occur naturally as a result of biological processes (e.g., methane). Most VOCs are a concern from a human health perspective and can also present aesthetic concerns.

VOC control is largely a concern in groundwater supplies as surface water supplies typically have lower concentrations of VOCs and are typically aerated from water sources by natural processes. VOCs can also be formed during the treatment process and throughout the transmission system. For example, trihalomethanes are formed when chlorine reacts with natural organic material in the water supply. Control of disinfection byproducts by aeration is discussed in this chapter.

VOCs most commonly encountered in groundwater supplies include halogenated hydrocarbons and VOCs associated with fuel and petroleum products. Chlorinated hydrocarbons include industrial solvents, degreasing, and cleaning agents. In addition, chlorinated hydrocarbons can be used or formed in the manufacturing process of other chemicals. Commonly encountered chlorinated hydrocarbons include, but are not limited to, carbon tetrachloride, dichlorobenzene, 1,2-dichloroethane (ethylene dichloride), 1,2-dichloroethylenes, ethylbenzene, methylene chloride, tetrachloroethylene (PCE), trichloroethylene (TCE), 1,1,1-trichloroethane, and vinyl chloride. A number of VOCs are used in fuel and petroleum products including but not limited to acetone, benzene, methyl tertiary butyl ether (MTBE), toluene, and xylenes.

A summary of commonly encountered VOCs and their potential sources is summarized in Table 4-7.

Applicable Regulations

VOCs in groundwater supplies include regulated VOCs and unregulated VOCs. The USEPA currently regulates 21 VOCs and has established maximum contaminant levels (MCLs) and maximum contaminant level goals (MCLGs) for each contaminant. The MCLG is a non-enforceable health goal and is set at a level in which there are no anticipated adverse health effects and allow for an adequate margin of safety (Ram 1990). For known human carcinogens, EPA sets the MCLG at zero because it is assumed that there is no known no-effect level (Ram 1990). The MCL is the enforceable legal limit for each contaminant and is set as close to the MCLG as possible based on available best technologies and cost (Ram 1990).

At a minimum, each state must promulgate standards as stringent as the federal regulations but can also adopt more stringent MCLs or regulate additional VOCs. New York State, for example, has established MCLs for 15 specific VOCs (1,1,1–trichloroethane, 1,1-dichloroethylene, 1,2,4-trichlorobenzene, chlorobenzene, cis-1,2-dichloroethylene, ethylbenzene, m-xylene, o-dichlorobenzene, o-xylene, p-dichlorobenzene, p-xylene, styrene, toluene, and trans-1,2-dichloroethylene). In addition, New York State regulates all unspecified organic contaminants with an MCL of 0.05 mg/L. Furthermore, California, Florida, and New Jersey have adopted a number of more stringent primary MCLs. California also adopts secondary MCLs based on taste and odor effects, for example, MTBE has an SMCL of 0.005 mg/L.

As discussed in Chapter 1, USEPA is developing a proposed carcinogenic VOCs (cVOC) rule. At the time of this writing, it is anticipated that MCLs for up to eight currently regulated VOCs may be lowered (benzene, carbon tetrachloride, 1,2-dichloroethane,

Table 4-7 Potential sources of commonly encountered VOCs (ATSDR)

VOC	Potential Sources
1,1,1-trichloroethane	Release/emissions from metal degreasing industry, emissions from manufacturing industry, paints, glues, and cleaning products
1,1,2-trichloroethane	Release from timber products, synthesis and plastics industry, laundries, paints, solvents, adhesives, and cleaning agents
1,1-dichloroethylene	Release from manufacturing process, polymers of chemical, proximity to inadequately controlled hazardous waste sites, and ingesting contaminated drinking water
1,2,4-trichlorobenzene	Release from manufacturing process, solvents, and dyes
1,2-dichloroethane	Release/emissions from cleaning agents, pesticides, glue, inhalation and ingestion of contaminated air, food, and drinking water
1,2-dichloropropane	Release from manufacturing process, solvents, proximity to inadequately controlled hazardous waste sites, and ingesting contaminated drinking water
Benzene	Inhalation of tobacco smoke and contaminated air in the vicinity of gas stations, and proximity to inadequately controlled hazardous waste sites
Carbon tetrachloride	Release from manufacturing process, ingestion of contaminated drinking water, and oral or dermal contact with contaminated soil
Chlorobenzene	Release from workplaces where chemical is used, proximity to inadequately controlled hazardous waste sites, inhalation and ingestion of contaminated air, food, and drinking water
Cis-1,2-dichloroethylene	Release from inadequately controlled hazardous waste sites and landfills, chemical spills, and breakdown of other chlorinated chemicals
Dichloromethane (Methylene Chloride)	Release from workplaces where chemical is used, paint, aerosol, workplaces where chemical is used, and proximity to inadequately controlled hazardous waste sites
Ethylbenzene	Emissions from vehicles, paint, solvents, pesticides, printing ink, varnishes, and inhalation of tobacco smoke
o-dichlorobenzene (1,2 –dichlorobenzene)	Release from herbicides, ingestion of contaminated drinking water
p-dichlorobenzene (1,4 –dichlorobenzene)	Release from workplaces where chemical is used, exposure to toilet-space deodorant and fumigant, inhalation and ingestion of contaminated air and drinking water
Styrene	Release from manufacturing process, insulation materials of buildings and electrical appliances, automobile parts, vehicle exhaust, tobacco smoke, plastic pipes, and fiberglass
Tetrachloroethylene	Release/emissions from metal degreasing industry and dry cleaning industry, and ingesting contaminated drinking water
Toluene	Release from adhesives, audio equipment, pharmaceutical and linoleum industries, from shoe and fabric manufacturers, and proximity to inadequately controlled hazardous waste sites
Trans-1,2-dichloroethylene	Release from inadequately controlled hazardous waste sites and landfills, chemical spills, and breakdown of other chlorinated chemicals
Trichloroethylene	Release from manufacturing process, paints, glues, metal cleaners, typewriter correction fluids, and ingesting contaminated drinking water
Total Xylenes (m-xylene, o-xylene, p-xylene)	Release from manufacturing process, biomedical laboratories, vehicle exhaust, paint, solvent, gasoline, rust preventatives, pesticides, and inhalation of tobacco smoke
Vinyl chloride	Inhalation of tobacco smoke, release from manufacturing process, proximity to inadequately controlled hazardous waste sites and landfills, and ingesting contaminated drinking water

1,2-dichloropropane, dichloromethane, tetrachloroethylene, trichloroethylene, and vinyl chloride). Furthermore, up to eight VOCs of the Third Contaminant Candidate List (CCL3) may also be regulated (1,3-butadiene, 1,1- dichloroethane, 1,2,3-trichloropropane, methyl oxirane, benzyl chloride, aniline, urethane, nitrobenzene).

A summary of relevant federal and state MCLs and MCLGs is provided in Table 4-8*.

Table 4-8 Relevant federal and state MCLs and MCLGs

Contaminant	Federal MCL (mg/L)	MCLG (mg/L)	NYS MCL (mg/L)	CA MCL (mg/L)	NJ MCL (mg/L)	FL MCL (mg/L)	Secondary MCLs
1,1,1,2 - tetrachloroethane			0.005				
1,1,2,2 - tetrachloroethane			0.005	0.001	0.001		
1,1,1-trichloroethane	0.2		0.005	0.2	0.03	0.2	
1,1,2-trichloro-1,2,2-trifluoroethane				1.2			
1,1,2-trichloroethane	0.005	0.003	0.005	0.005	0.003	0.005	
1,1-dichloroethane			0.005	0.005	0.05		
1,1-dichloroethylene	0.007		0.005	0.006	0.002	0.007	
1,1-dichloropropene			0.005				
1,2,3 -trichlorobenzene			0.005				
1,2,4 -trichlorobenzene	0.07	0.07	0.005	0.005	0.009	0.07	
1,2,4 -trimethylbenzene			0.005				
1,2-dichloroethane	0.005		0.005	0.0005	0.002	0.003	
1,2-dichloropropane	0.005		0.005	0.005	0.005	0.005	
1,3,5 -trimethylbenzene			0.005				
1,3-dichloropropane			0.005				
1,3-dichloropropene				0.0005			
2,2 -dichloropropane			0.005				
2 -chlorotoluene			0.005				
4 -chlorotoluene			0.005				
Benzene	0.005		0.005	0.001	0.001	0.001	
Bromobenzene			0.005				
Bromochloromethane			0.005				
Bromomethane			0.005				
Carbon tetrachloride	0.005		0.005	0.0005	0.002	0.003	
Chlorobenzene (Monochlorobenzene)	0.1	0.1	0.005	0.07	0.05	0.1	
Chloroethane			0.005				
Chloromethane			0.005				
Cis-1,2-dichloroethylene	0.07	0.07	0.005	0.006	0.07	0.07	
Cis-1,3-dichloropropene			0.005				
Dibromomethane			0.005				
Dichlorodifluoromethane			0.005				
Dichloromethane (Methylene Chloride)	0.005	0	0.005	0.005	0.005	0.005	

* Maximum contaminant levels are current as of Sept. 20, 2013.

Continued

Table 4-8 Relevant federal and state MCLs and MCLGs *(continued)*

Contaminant	Federal MCL (mg/L)	MCLG (mg/L)	NYS MCL (mg/L)	CA MCL (mg/L)	NJ MCL (mg/L)	FL MCL (mg/L)	Secondary MCLs
Ethylbenzene	0.7	0.7	0.005	0.3	0.7	0.7	
Hexachlorobutadiene			0.005				
Isopropylbenzene			0.005				
m-dichlorobenzene (1,3-dichlorobenzene)			0.005		0.6		
Methyl tert-butyl ether (MTBE)			0.01	0.013	0.07		CA - 0.005
m-xylene	10.0		0.005	1.75	1.0	10.0	
Naphthalene					0.3		
n-butylbenzene			0.005				
n -propylbenzene			0.005				
o-dichlorobenzene (1,2-dichlorobenzene)	0.6		0.005	0.6	0.6	0.6	
o-xylene	10.0		0.005	1.75	1.0	10.0	
p-dichlorobenzene (1,4-dichlorobenzene)	0.075		0.005	0.005	0.005	0.075	
p -isopropyltoluene			0.005				
p-xylene	10.0		0.005	1.75	1.0	10.0	
sec-butylbenzene			0.005				
Styrene	0.1		0.005	0.1	0.1	0.1	
tert-butylbenzene			0.005				
Tetrachloroethylene (PCE)	0.005	0	0.005	0.005	0.001	0.003	
Toluene	1.0	1.0	0.005	0.15	1.0	1.0	
Trans-1,2-dichloroethylene	0.1	0.1	0.005	0.01	0.1	0.1	
Trans-1,3-dichloropropene			0.005				
Trichloroethylene (TCE)	0.005		0.005	0.005	0.001	0.003	
Trichlorofluoromethane			0.005	0.15			
Total Xylenes	10.0	10.0	0.005	1.75	1.0	10.0	
Vinyl chloride	0.002		0.002	0.0005	0.002	0.001	

CA—California, FL—Florida, MCL—maximum contaminant level, MCLG—maximum contaminant level goal, NYS—New York State

Important VOC Physical Properties

Important VOC physical properties include Henry's law constant, solubility, vapor pressure, specific gravity, and air and water diffusion coefficients. Each of these parameters is discussed in further detail in Chapter 2.

Henry's law constants for commonly encountered VOCs are summarized in Table 4-9. The Henry's law constants have been grouped based on affinity for air stripping as follows:

- $K_H > 0.05$: VOC is amenable to air stripping
- $0.05 > K_H > 0.002$: VOC is moderately amenable to air stripping
- $K_H < 0.002$: VOC is not amenable to air stripping

Table 4-9 Range of Henry's law constants (dimensionless) for select VOCs

K_H Range*	VOC
$K_H > 0.05$	1,1,2-trichloro-1,2,2-trifluoroethane[†]
	1,1,1-trichloroethane
	1,1,2-trichloroethane
	1,1-dichloroethane
	1,1-dichloroethylene
	1,2,3 -trichloropropane[†]
	1,2,4 -trichlorobenzene
	1,2 -dichloroethane
	1,2 -dichloropropane
	1,3,5 -trimethylbenzene
	Benzene
	Carbon tetrachloride
	Chloroethane
	Chlorobenzene (monochlorobenzene)
	Cis-1,2-dichloroethylene
	Dichloromethane (methylene chloride)
	Ethyl benzene
	m-dichlorobenzene (1,3-dichlorobenzene)
	m-xylene
	Naphthalene[†]
	o-chlorotoluene[†]
	o-dichlorobenzene (1,2–dichlorobenzene)
	o-xylene
	p-dichlorobenzene (1,4–dichlorobenzene)
	p-xylene
	Styrene
	Tetrachloroethylene
	Toluene
	Trans-1,2-dichloroethylene
	Trichloroethylene
	Trichlorofluoromethane
	Vinyl chloride
$0.05 > K_H > 0.002$	1,1,1,2 – tetrachloroethane[†]
	1,1,2,2 - tetrachloroethane
	1,1,2-trichloroethane
	1,2,3 –trichlorobenzene[†]
	1,2,4 –trimethylbenzene[†]
	1,3 –dichloropropane[†]
	1,3 –dichloropropene[†]
	Bromobenzene[†]
	Bromomethane[†]
	Chloromethane[†]

Continued

Table 4-9 Range of Henry's law constants (dimensionless) for select VOCs *(continued)*

K_H Range*	VOC
	Cis-1,3-dichloropropene†
	Dibromomethane†
	Hexachlorobutadiene†
	Isopropylbenzene†
	MTBE
	sec-butylbenzene†
	Styrene†
	tert-butylbenzene†
	Trans-1,3-dichloropropene†
	Bromochloromethane†
$K_H < 0.002$	Acetone
	Dichlorodifluoromethane†
	Tertiary butyl alcohol
	1,2,3-Trichloropropane (TCP)

VOC—volatile organic compound
* Refer to Table 2-3 in Chapter 2 for a detailed list of Henry's Law Constants reported in literature
† Reference requires verification

A more detailed list of Henry's law constants is also provided in Table 2-3 of Chapter 2.

Air stripping can be used to remove regulated and unregulated VOCs from the water supply to achieve compliance with MCLs and/or utility-specific finished water treatment criteria. Air stripping can either be used alone to meet finished water design criteria or can be used in combination with another VOC treatment technology to achieve desired treatment objectives. For example, air stripping can be used in combination with granular activated carbon (GAC) polishing in order to minimize carbon replacement costs and treat a wide range of VOC compounds.

Primary Removal Technologies

The EPA has identified packed tower aeration and GAC as "best available technology" (BAT) for VOC removal (Ram 1990). Principal technologies used for air stripping VOCs include packed tower aeration, low-profile aerators, diffused bubble aeration systems, and surface aeration. Refer to Chapter 2 for a more detailed discussion on these aeration technologies. Other aeration technologies that could be considered include hydraulic jet venturis, vacuum air strippers, and membrane degassifiers. However, these technologies are not typically used for drinking water applications. Selecting the type of aeration technology for a specific application is typically based on contaminant type and concentration, flow rate, site constraints, and costs.

Quantifying the nature and extent of groundwater contamination can be challenging due to limited data, cost of monitoring, inaccessibility to groundwater, and/or inability to collect samples that are representative of full-scale pumping conditions (Ram 1990). This uncertainty in VOCs (both in type and concentrations) must be addressed in the design of aeration equipment. Available water quality data and groundwater fate and transport modeling can be used (where available) to estimate future trends in VOC concentrations.

However, a safety factor is typically applied to raw water design concentrations to account for potential variability in the water supply. Safety factors can range from two to five or greater depending on the extent of data available, confidence in the data, and criticality of the supply (see Chapter 2).

Air Quality Considerations

Regulatory context and basis of compliance. The process of aeration as employed in drinking water treatment is intended to encourage mass transfer of an undesirable substance from the liquid (dissolved) phase, to the gaseous phase. The mass of the contaminant is conserved, albeit often at a diluted gaseous concentration. Put simply, a treatment process involving aeration transfers a contaminant out of the liquid process stream, and into a gaseous waste stream of air.

The ambient atmosphere typically serves as both source and receiving reservoir for the process air stream in the act of aeration, and therefore some consideration must be given to the quality (i.e., the chemical composition) of air discharged as waste. The atmospheric air of the United States is regulated under the Clean Air Act (CAA). The CAA was originally passed in 1963, establishing funding for research regarding air pollution. Amendments were passed in 1970, 1977, and 1990—each strengthening and expanding regulatory purview with regard to air quality, the breadth of regulated contaminants, and enforcement powers of USEPA.

The CAA was incorporated into the United States Code as Title 42, Chapter 85. In its most current format, the CAA empowers USEPA to regulate the following contaminant types:

- National Ambient Air Quality Standards (NAAQS)—specifically, six (6) common air pollutants:
  - ground level ozone
  - particulate matter
  - carbon monoxide
  - nitrogen oxides
  - sulfur dioxide
  - lead
- ozone layer protection
- toxic air pollutants
- visibility

Water treatment professionals are typically most concerned with the broad category of contaminants entitled "Toxic Air Pollutants." This classification provides governance – in terms of concentration and allowable mass of discharge—for a list of 187 airborne contaminants known to cause cancer or other serious health effects. A list of these contaminants is given in this chapter in The Clean Air Act Amendments of 1990 List of Hazardous Air Pollutants (2013b). Additional information is included in Appendix A.

According to USEPA (2013a), most air toxics originate from human-made sources, including mobile sources (e.g., cars, trucks, buses) and stationary sources (e.g., factories, refineries, power plants), as well as indoor sources (e.g., building materials and activities such as cleaning). However, there are two types of stationary sources that generate routine emissions of air toxics:

1. "Major" sources are defined as sources that emit 10 tons per year of any single species of the listed toxic air pollutants, or 25 tons per year of a mixture of air toxics. These sources may release air toxics from equipment leaks, when materials

are transferred from one location to another, or during discharge through emission stacks or vents

2. "Area" sources consist of smaller-size facilities that release lesser quantities of toxic pollutants into the air. Area sources are defined as sources that emit less than 10 tons per year of a single air toxic, or less than 25 tons per year of a combination of air toxics. Though emissions from individual area sources are often relatively small, collectively their emissions can be of concern—particularly where large numbers of sources are located in heavily populated areas.

Based on these definitions, most water treatment facilities utilizing aeration to remove a VOC or dissolved gas do not qualify to be listed as Major Sources, but more typically meet the criteria of an Area Source. As an example, take a single groundwater source of capacity 1.0 MGD, treating a dissolved VOC at a hypothetical concentration of 500 ppb (μg/L), or 0.5 ppm (mg/L). Assuming complete transfer of the VOC and year-round operation, the facility would discharge a maximum mass of 4.17 lb/day, or approximately 1,522 lb/year, based on the following equation:

$$8.34\, C_{VOC} Q = M_{VOC} \qquad \text{(Eq 4-11)}$$

where:

C_{VOC} = VOC concentration in mg/L
Q = Liquid process flow rate in MGD
M_{VOC} = Mass of contaminant removed in lb/day

Note that the 8.34 coefficient is a conversion factor with units of lbs/(MGD·mg/L). Observing the 10-ton definition for a "Major Source," and again assuming 100 percent contaminant transfer during aeration, a solution can be reached for the Major Source threshold contaminant concentration at a given facility. Solving this equation for C_{VOC} we note that a contaminant concentration of 6,570 ppb (μg/L) per MGD is the minimum required to meet the EPA Major Source criteria.

Permitting and compliance. USEPA typically delegates primacy for CAA promulgation, compliance, and enforcement responsibilities to state-level environmental protection agencies or departments. The CAA itself makes no provision for compliance or enforcement for non-major sources. However, some state agencies have instituted stricter standards. As such, any utility or Public Water System (PWS) operator considering an aeration process should first perform compliance due diligence with the regulatory agency holding primacy over that PWS service area (USEPA 1998).

Some state-level agencies may add chemical species to the list currently categorized as Toxic Air Pollutants by USEPA. Additionally, state agencies may mandate individual species as being subject to special regulation or may designate distinct categories of air toxics for the purpose of establishing permit threshold concentrations, in either a waste air stream or in a liquid process stream undergoing treatment. The greater point to understand is that USEPA establishes a baseline for the regulation and monitoring of air toxics. The state-level agencies (to which USEPA delegates regulatory primacy) may, and often do, choose to implement the baseline federal regulations in a more stringent or detailed manner, with the potential for additional permitting, monitoring, and enforcement requirements.

Specific to the operation of an aerator or air stripper, it is important to note that some states currently regulate water treatment facilities with waste air streams that would not qualify as "Major Sources," with regard to the 10-ton annual limit for discharge of VOCs. As an example of additional regulation (NJAC 2013), the State of New Jersey currently regulates water treatment aeration facilities via the following code language:

Equipment which is used for treating groundwater, industrial waste water, or municipal wastewater with a solids content of less than two percent by weight as it enters the equipment (typical operations performed by this type of equipment include, but are not limited to, air stripping, aeration, digestion, thickening, flocculating, surface impounding, and dewatering), if the equipment does either of the following:

(i) Treats or handles influent which has one or both of the following:

(1) A total concentration of VOCs and Group 2 TXS in the influent of 3,500 parts per billion by weight (ppbw) or more; or

(2) A total Group 1 TXS concentration in the influent of 100 ppbw or more;

This language is relevant to note as an example, since it depicts the fact that a state-level agency with primacy may enforce regulatory requirements beyond the CAA, based on one of the following criteria:

- Designation of special groups of air toxic contaminants: In New Jersey's example code language, several chemical species are designated as belonging to "Group 1 TXS" and "Group 2 TXS," indicating special categories of toxicity, and having originated from the baseline group of 187 chemical species listed under the Toxic Air Pollutants category of the CAA).
- Designation of liquid process stream concentrations as regulatory criteria and/ or permitting triggers, at values lower than might otherwise be required under the CAA: In New Jersey's example, a process stream threshold of 3,500 ppb is established for regulation of a water treatment facility by permit. This criterion is independent of the CAA annual limit of 10 tons for discharge of a single VOC species, and therefore independent of facility liquid process stream flow rate.

The above data are given for example only; however, they serve to illustrate the fact that individual system operators must consult their own state-level regulators to determine what—if any— additional criteria may apply to their aeration facilities, beyond the "Major Source" threshold as established by EPA. In a second example, California enforces risk-based requirements that limit emissions from new permits without imposing emissions control measures. The risks include limits for maximum individual cancer risk (MICR), cancer burden, and noncancer acute and chronic hazard index (HI), based on nearby residential and commercial receptors. Note finally that some states (e.g., California) may require separate permits for construction and for operation of treatment equipment related to indoor and/or atmospheric air quality.

In addition to the determination of permitting requirements for construction, an operating permit related to any process stream, whether in the liquid or gas phase, will typically require some form of monitoring, analysis, and scheduled reporting–potentially by a certified environmental laboratory service. If not equipped to handle this specialized form of monitoring in-house, operators are advised to solicit the services of a NELAC-certified environmental laboratory, with expertise in ambient air quality assessment and monitoring. It is preferable that such lab services provide for their own testing media as part of their standard monitoring contract, in accordance with the terms and conditions of a given facility's monitoring requirements.

Aesthetic concerns, health effects, and dispersion modeling. A review of regulatory criteria regarding air quality monitoring indicates that if a facility neither meets USEPA criteria of Major Source, nor falls under any state-level permit or monitoring scheme, then no treatment is required for the waste gas stream. However, water treatment facility operators may wish to consider the following additional criteria when evaluating the potential value of off-gas treatment:

- Location of treatment facility
- Potential for movement of the off-gas air plume

- Threshold concentrations for health effects of airborne chemical species

USEPA publishes two tables indicating human health risk based on both acute and chronic modes of exposure to airborne toxic pollutants. These tables are Table A-1 and Table A-2 in Appendix A.

Aside from the due diligence required to ascertain regulatory compliance requirements for an aeration facility, an operator should consider a chemical assay of the facility's process fluid to ascertain presence and concentration of the contaminants listed in Tables A-1 and A-2 in Appendix A. These contaminants are recognized by USEPA as creating a negative health effect in human beings. If the initial assay (evaluated in tandem with treatment process calculations) indicates the potential for a detectable concentration of one or more of the listed contaminants in the facility waste air stream—the owner/operator may consider taking steps to better understand the atmospheric dispersion of his or her air waste stream.

Such mitigating steps may include the following:

- Performance of bench-scale, pilot-scale, or full-scale stack testing: Characterize waste air stream contaminant species and concentrations via empirical data.
- Dispersion analysis: Evaluation of the atmospheric dispersion of waste air stream plume, including likely mixing and dilution characteristics in ambient air, and prediction of likely ambient contaminant concentrations at neighboring properties. Note that many state regulatory agencies provide free software for dispersion modeling. Proprietary software is also available for purchase. Alternately, a consulting scientist or engineer may be retained to perform such an analysis.

Regulatory compliance summary. In summary, the owner/operator of a potable water treatment works must consider two major criteria when evaluating the regulatory consequences of an aeration or air stripping unit process:

1. EPA and State-Level Permit Criteria – Evaluate raw water contaminant characteristics, and determine if either the process fluid concentration, or the predicted mass of annual off-gas contaminants meet the criteria for a USEPA-defined Major Source. Evaluate further based on the permitting criteria of the state-level environmental agency with CAA primacy at the location of the proposed treatment facility. Define whether separate permits will be required for construction and operation and obtain a written statement from the environmental regulatory agency regarding type of reporting documentation required, a listing of laboratories certified to perform sampling and analysis, and the reporting frequency required for the facility in question.
2. Human Health Criteria – Perform sufficient testing to determine the potential chemical concentrations of any contaminants being transferred to a plant's air waste stream. Perform additional testing to determine the extent of exposure to neighboring properties downwind of the emission stack, packed tower vent, or chimney stack, in terms of frequency and concentration of exposure. Compare the results of Criteria No. 1 to USEPA regulations published under the CAA, and state-level regulations governing the location of the proposed treatment unit.

Compare the results of Criteria No. 2 to USEPA human health exposure thresholds (both acute and chronic) and determine residents or employees at neighboring properties may be subject to concentrations beyond the listed thresholds. If any of the listed contaminants are present above a concentration of "non-detect," consider the use of safety factors in determining whether waste air stream concentration is warranted; specifically, consider the value of treating waste air streams which present detectable levels of contaminant(s), but may not meet the criteria of acute and/or chronic exposure health risk. Finally, in evaluation of the consequences of treatment vs. no action for waste stream air treatment at facilities with a known toxic air pollutant, retain technical and/or legal counsel with environmental practice expertise, to help characterize and weigh the risks of undertaking voluntary, non-compulsory treatment steps. Consider risk of exposure and compare against capital and operating costs for treatment equipment.

The results of these comparisons will inform an operator regarding the regulatory requirements and health-risk consequences associated with the construction and operation of an aeration or air stripping treatment unit. If gas-phase treatment is considered to be warranted, another decision-making process is required to determine the most effective and economic format for such treatment. Options for gas-phase waste stream treatment, and an approach to unit process decision-making, are outlined below.

OFF-GAS TREATMENT AND RESIDUALS MANAGEMENT FOR AERATION SYSTEMS

Options for treating off-gases from aeration or air stripping processes include the following:

- Aeration system design modification
- Vapor phase granular activated carbon (GAC) adsorption
- Thermal oxidation
- Catalytic oxidation
- Other (emerging) technologies

Modification of the aeration system may be feasible where the air emission standard is based on projected ground level concentrations (mg/m^2). In such cases, increasing the stack height or the air flow rate may be cost-effective measures for dispersing the VOCs in the off-gas.

Of the technologies available for the treatment of the off-gas from a drinking water treatment plant engineered for VOC removal, the most typical is the use of granular activated carbon (GAC) adsorption contactors. Applicability varies by installation, but generally GAC units have proven to be the most cost-effective methodology for achieving EPA and state-level emissions criteria for ambient air quality (Crittenden et al. 1988).

Vapor Phase GAC Adsorption

The treatment process for GAC-based treatment of a waste air stream is illustrated in Figure 4-7.

The liquid process stream is routed through an aeration unit, most typically a packed tower aerator. The focus of Figure 4-7, however, is the routing and conditioning of the VOC-laden waste air stream:

1. Clean air is routed through the packed tower, via blower, counter-current to the liquid stream.
2. The waste air stream is heated to reduce relative humidity to below 50 percent (40–50 percent is considered optimal).
3. Waste air is then routed through a gas-phase GAC contactor, typically at a design velocity of 60 ft/min (100 ft/min maximum).

Figure 4-7 also shows an on-site steam regeneration option that recovers the VOCs and allows the GAC to be reused after it reaches its adsorptive capacity. Operators and designers must determine whether on-site regeneration is appropriate and cost-effective for a given site, weighing operator intensive regeneration processes against the cost of replacing GAC. In most cases for drinking water systems, single use of the vapor phase GAC is preferred.

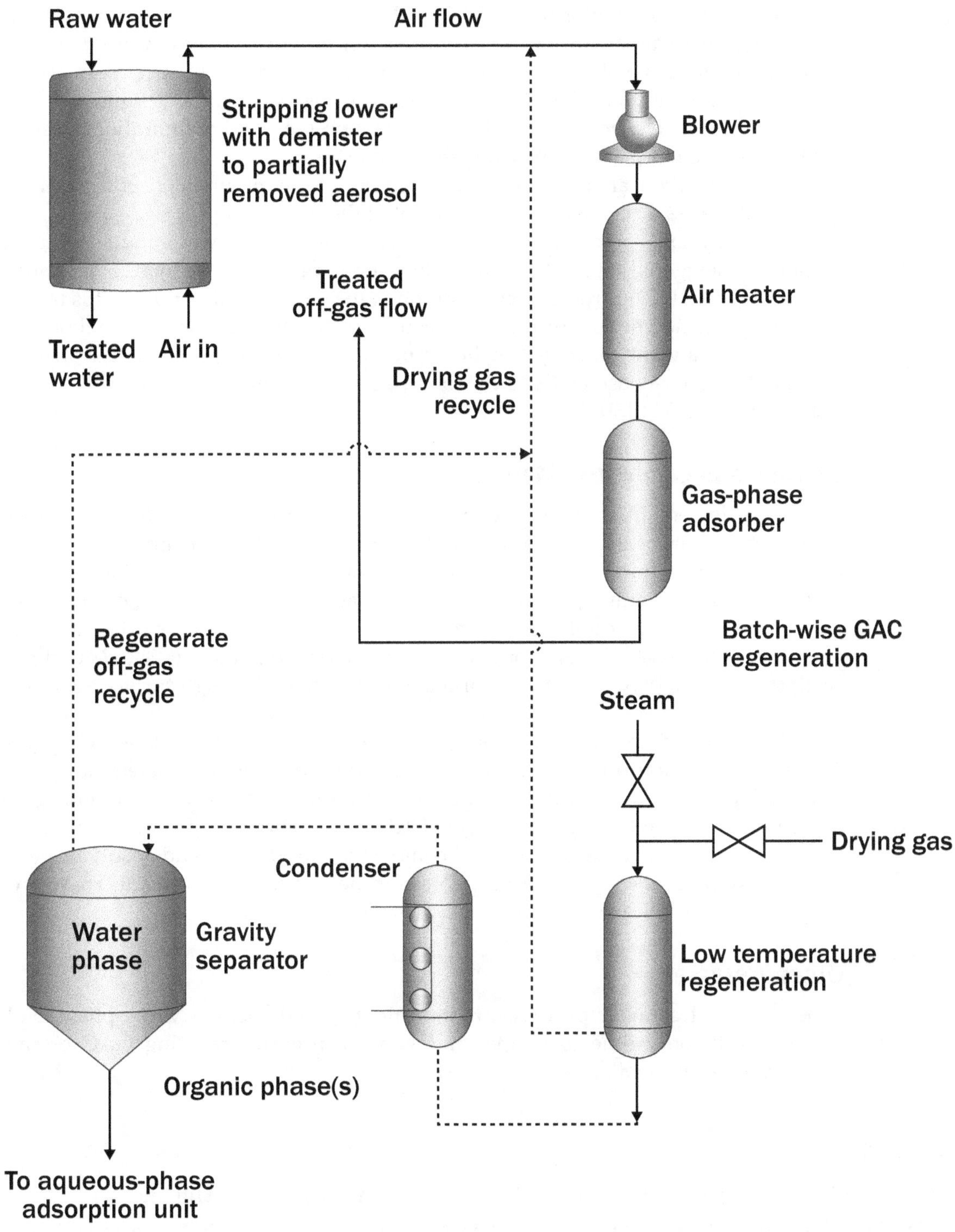

Figure 4-7 Treatment scheme for removing VOCs from air stripper off-gas

An important characteristic of GAC is the quantity of adsorbate (contaminant chemical) that it can accumulate. The adsorption isotherm describes the equilibrium relationship between the adsorbate, adsorbent, and the equilibrium concentration of the adsorbate in [fluid]. GAC isotherms are typically shown graphically on log-log plots. On such plots, more adsorbable compounds have higher and flatter lines than less adsorbable compounds (USEPA 2013c).

An example of a vapor phase GAC isotherm is given here (Chowdhury 2013):

In a given treatment design scenario, a catalog of GAC isotherms is used to choose the GAC species most suitable for adsorption of a given contaminant species. In the simplest and most frequently employed design scenario, the designer will build his treatment process around the concept of fixed-bed adsorption (i.e., the configuration where a GAC media is isolated in-place as an adsorptive reactor).

GAC adsorbers are commonly designed in redundant, parallel configurations, with or without the GAC regeneration component of the treatment process.

Typically, one or more units are operated in flow-forward mode, with batches initiated at staggered intervals, while at least one unit is kept on standby. As any single unit approaches breakthrough concentration at the effluent point, that unit is taken out of service for regeneration or media changeout, and the standby unit put online, ensuring continuous treatment capacity. Detailed guidance on the design of GAC systems for the treatment of vapor-phase VOCs can be found elsewhere (Chowdhury et al. 2013, Crittenden et al. 1998, USEPA 2013b).

Thermal and Catalytic Oxidation

Thermal oxidizers are devices in which the air stream containing VOCs is passed over or through a burner flame or other pre-heated device into a chamber where the VOCs are oxidized. The units are typically single-chamber, refractory-lined oxidizers equipped with a propane or natural gas burner and an exhaust stack. In contrast, catalytic oxidizers incorporate a bed of catalytic surfaces to initiate and promote oxidation at a lower temperature (600–800°F) than would be used in thermal oxidation (1,350–1,800°F). Typical catalysts are composed of a ceramic or metal substrate with a high surface area to volume ratio.

Thermal and catalytic processes can destroy over 99 percent of VOC emissions. The major oxidation products produced are carbon dioxide and water. However, other undesirable combustion byproducts including carbon monoxide, oxides of nitrogen (NO_x) and sulfur (SO_x), and halogenated acids HCl, HF, HBr) can be produced, depending on the VOCs present. These undesirable combustion products could result in permitting problems or present corrosive atmospheres for the treatment equipment. These processes are potentially applicable to non-halogenated VOCs.

Other Processes

Other off-gas treatment processes that could be used include vapor-phase biological aerobic oxidation, ozone oxidation, and vapor concentrators using GAC or modified zeolite adsorptive media.

REFERENCES

AWWA. 1999 (5th ed.). *Water Quality and Treatment.* New York: McGraw-Hill, Inc.

AWWA and ASCE. 1990 (3rd ed.). *Water Treatment Plant Design.* New York: McGraw-Hill, Inc.

Blute, N.; Ghosh, A.; and Lytle, D. 2012. Assessing Ammonia Treatment Options. *Opflow,* 38(5).

Böhlke, J.K.; Smith, R.L.; and Miller, D.N. 2006. Ammonium Transport and Reaction in Contaminated Groundwater: Application of isotope tracers and isotope fractionation studies, Water Resources Research, Vol. 42, W05411.

Brooke, E. and M. R. Collins. 2011. Posttreatment Aeration to Reduce THMs. *Journal American AWWA,* 103(10):84–96.

Chowdhury, Z.K., et al. 2013. Activated Carbon: Solutions for Improving Water Quality. Denver, CO: AWWA.

Clark, T.F. 2013. "THM Removal: Small System Application." Presentation at the Finger Lakes Conference, Watkins Glen, New York.

Clegg, S. and Whitfield, M., 1995. A chemical model of seawater including dissolved ammonia and the stoichiometric dissociation constant of ammonia in estuarine water and seawater from -2 to 40°C. *Geochim. Cosmochim. Acta,* 59(12).

Corbitt, R.A. 1990. Standard Handbook of Environmental Engineering. New York: McGraw-Hill.

Crittenden, J.C.; Cartright, R.D.; Rick, B.; Shin-Ru Tang, and Perram, D. 1988. Using GAC to Remove VOC's from Air Stripper Off-Gas. *Journal. AWWA,* 80:5:73-84.

Crittenden, J.C.; Trussell, R.R.; Hand, D.W.; Howe, K.J.; and Tchobanoglous, G. 2012. *MWH's Water Treatment: Principles and Design*. New York: John Wiley & Sons, Inc.

Cummings, M.D. 1988. *Removal of Radon from Contaminated Ground Water by Packed Column Air Stripping, Blairsville, GA.* Cincinnati, Ohio: USEPA Office of Drinking Water, Technical Support Division.

Deb, A.K. 1992. *Contribution of Waterborne Radon to Home Air Quality.* Denver, CO: Awwa Research Foundation and AWWA.

Dixon, K.L.; Lee, R.G.; Smith, J.; and Zielinski, P. 1991. Evaluating Aeration Technology for Radon Removal. *Journal. AWWA,* 83(4):151–148.

Dohnalek, D.A. and J.A. FitzPatrick. 1983. The chemistry of reduced sulfur species and their removal from groundwater supplies. *Journal AWWA,* 75(6): 298–308.

Drago, J.A. 1998. *Critical Assessment of Radon Removal Systems for Drinking Water Supplies.* Denver, CO: Awwa Research Foundation and AWWA.

Duranceau, S.J., et. al. 2010. Innovative Hydrogen Sulfide Treatment Methods: Moving Beyond Packed Tower Aeration. *Florida Water Resources Journal.*

Duranceau, S.J.; Anderson, R.K.; and Teegarden, R.D. 1999. Comparison of Mineral Acid Pretreatments for Sulfide Removal. *Water Utility Operations.* Denver, CO: American Water Works Association.

Fiske, P. S.; Oppenheimer., R. J.; Moore, R.; and Everett, R. 2011. In-Tank Aeration Predicts and Reduces THMs. *AWWA Opflow,* 22–24.

Flinn, A.D.; Weston, R.S.; and Bogert, C.L. 1916. *Waterworks Handbook. New York:* McGraw-Hill Book Company Inc.

Florida Administrative Code, Chapter 62-555, Permitting, Construction, Operation, and Maintenance of Public Water Systems.

Ghosh, A.; Seidel, C.; Townsend, E.; Pacheco, R.; and Corwin, C. 2015. "Reducing Volatile Disinfection Byproducts in Treated Water Using Aeration Technologies," Water Research Foundation Project 4441 Final Report, Denver, CO.

Hawthorn, J.E. 1970. Hydrogen Sulfide Damage to Concrete Pipe. *Journal WPCF,* (42):425.

Haynes, W.M. (editor). 2014 (95th ed.). *CRC Handbook of Chemistry and Physics.* Boca Raton, FL: CRC Press LLC.

Henrie, T.; Rezania, L.-I. W.; Nagy, D.; and Lytle, D. 2012. Got Ammonia? *Opflow,* 38(6).

Hiltebrand, D.J.; Dyksen, J.E.; and Raman, K. 1987. Radon in Water Supply Wells: Treatment Facility Requirements and Costs. In Proceedings of the NWWA Conference, Somerset N.J., April 7-10, 1987. Chelsea, MI: Lewis Publishers and NWWA.

Hirschhorn, E.A. and R. Moore. 2014. Reduction of THMs within Storage Tanks Using Aeration. *Proceedings,* Pacific Northwest Section AWWA, 2014 Annual Conference.

History of Hydrogen Sulfide Ambient Air Quality Standards. 2009. California Environmental Protection Agency Air Resources Board. http://www.arb.ca.gov/ research/aaqs/caaqs/ h2s/h2s.htm.

Jacobs, S.; Reiber, S.; and Edwards, M. 1998. Sulfide-Induced Copper Corrosion. *Journal AWWA*, 90(7):62.

Jacobsen, L.; Fang, M.; and Machado, D. 2011. "Aeration System Addition and Operation for Distribution System Trihalomethane Control," Proc. AWWA Ann. Conf. Washington, DC.

Kansas Department of Health and Environment. 2010. Guidelines for Investigating and Remediating Nitrate/Ammonia Contamination from Agricultural Chemical Releases, Policy BER-RS-050, KS.

Kavanaugh, Michael C. and R. Rhodes Trussell. 1980. Design of Aeration Towers to Strip Volatile Contaminants from Drinking Water. American Water Works Association: Research and Technology.

Kibbel, W.H. 1976. Peroxide Treatment for Industrial Waste Problems. *Indus. Wtr. Egnrg.*, 13(4):6.

Kinner, N.E.; Malley Jr., J.P.; Clement, J.A.; Quern, P.A.; and Schell, G.S. 1990. *Radon Removal Techniques for Small Community Public Water Supplies.* EPA 600/S2-90/036. Springfield, Va.: National Technical Information Service. PB 90-257809/AS.

Kinner, N.E.; Malley, J.P.; Clement, J.A.; and Fox, K.R. 1993. POE Techniques to Remove Radon. *Journal AWWA*, 85(6):75–86.

Kinner, N.E.; Quern, P.A.; Schell, G.S.; Lessard, C.E.; and Clement, J.A. 1089. *Treatment Technology for Removing Radon from Small Community Water Supplies.* EPA 600/0-89-249. Washington, D.C.: USEPA.

Landolt-Boernstein. 1976. Zahelenwerte und Funktionen aus Physik u.s.w., Vol. IV. Part 4, Band c. Berlin and New York: Springer-Verlag.

Lowry, J. 1988. Point-of-Entry Radon Removal by GAC and Aeration. In *Proceedings of the Third National Conference on Drinking Water. Drinking Water Treatment. Small System Alternatives.* (P. Toft, R.S. Tobin, and J. Sharp, editors). Oxford: Pergamon Press.

Lowry, J.D. 2010. "Removal of Sulfides Using Carbon Dioxide and Multi-Staged Bubble Air Stripping," Proceedings of the 2010 Inorganic Contaminants Workshop, American Water Works Association, Denver, February 28 – March 2.

Lowry, J.D. and J.E. Brandow. 1985. Removal of Radon from Water Supplies. *Journal of Environmental Engineering*, 111(4):511–525.

Lowry, J.D. and S.B. Lowry. 1988. Radionuclides in Drinking Water. *Journal AWWA*, 80(7):51–64.

Lowry, J.D.; Brutsaert, W.F.; McEnery, T.; and Molk, C. 1984. GAC Adsorption and Diffused Aeration for the Removal of Radon from Water Supplies. In *Proceedings of the AWWA 1984 Annual Conference.* Denver, CO: AWWA.

Lyn, T. and J. Taylor. 1992. Assessing Sulfur Turbidity Formation Following Chlorination of Hydrogen Sulfide in Groundwater. *American Water Works Association Research and Technology Journal.*

Lyn, T.; Taylor, J.; and Powell, R. 1993. *Turbidity Formation During Hydrogen Sulfide Chlorination.* Miami, FL: AWWA WQTC.

McKone, T.E. 1987. Human Exposure to Volatile Organic Compounds in Household Tap Water: The Indoor Inhalation Pathway. *Environ. Sci. Technol.*, 21(12):1194-1201.

Narasimhan, R.; Lowry, J.D.; Culley, J.; and Young-Pong, N. 2005. Management of the Disposal of Radioactive Residuals in Drinking Water Treatment, AwwaRF Final Report for Project No. 2695.

NJAC 2013. Permits and Certificates for Minor Facilities (and Major facilities without an Operating Permit). New Jersey Administrative Code, Title 7, Chapter Subchapter 8, Section 7:27-8.2 Applicability. Excerpted.

Odell, L. H. 2010. *Treatment Technologies for Groundwater*. Denver, CO: AWWA.

Office of Environmental Health Hazard Assessment (OEHAA). 2013. http://oehha.ca.gov/air/chronic_rels/pdf/7783064.pdf.

Overbeck, Paul K. 1995. Ground Water Color and Sulfide Reduction with Ozone. *AWWA ACE.* Anaheim, CA.

Pedersen, O.; Colmer, T.D.; and Sand-Jensen, K. 2013. "Underwater Photosynthesis of Submerged Plants – Recent Advances and Methods." *Frontiers in Plant Science*. 4.

Perkins, W.D. and J.C. Hewett. 1988. Full Scale Removal of Radon Gas by Diffused Aeration in Steep Falls, Maine. In *Proceedings of the 1998 AWWA Annual Conference.* Denver, CO: AWWA.

Rafferty, J.C. 1863. Some Notes on a Plant for the Removal of Radon by Aeration. *Jour. Institution Wat. Engrs.*,17: 23-25.

Ram, N.M.; Christman, R.F.; and Cantor, K.P. 1990. *Significance and Treatment of Volatile Organic Compounds in Water Supplies.* MI: Lewis Publishers, Inc.

Reid, L.W. 2012. Localized Treatment Solutions for TTHM Management. *New York Section AWWA Annual Water Event & Expo 2012*. New York.

Roberts, P.V. and J.A. Levy. 1985. Energy Requirements for Air Stripping Trihalomethanes. *Journal AWWA*, 77(4):138–146.

Schneider, O.D.; LeChevallier, M.W.; Yang, J.; Hughes, D.M.; and Reed, H. 2015. "Localized Control of Disinfection By-Products by Spray Stripping in Storage Tanks". *Water Research Foundation Project 4413 Final Report*, Denver, CO.

Seidel, C.J.; Acquafredda, S.; Jensen, J.M.; and Tang, G. 2010. Aeration Strategies for Total Trihalomethanes (TTHMs) Control, *Proceedings, AWWA Annual Conference,* Chicago.

Sherant, S. R. 2008. *Trihalomethane Control by Aeration.* Masters Thesis : The Pensylvania State University.

Siekkinen, A.R.; Mclellan, J.M.; Chen, J.; and Xia, Y. 2006. Rapid Synthesis of Small Silver Nanocubes by Mediating Polyol Reduction with a Trace Amount of Sodium Sulfide or Sodium Hydrosulfide. *Chemical Physics Letters*, 432(4–6):491.

Suffet, I. H. 1995. *Advances in Taste-And-Odor Treatment and Control*. Denver, CO: American Water Works Association Research Foundation.

Toccalino, P.L. and J.A. Hopple. 2010. Quality of Water from Public Supply Wells in the United States 1993-2007, Circular 1346, United States Geological Survey, Reston, VA.

US Department of Health, Education, and Welfare (USDHEW). 1970. Radiological Health Handbook. Washington, D.C.: U.S. Government Printing Office.

United States Department of Health and Human Services, Public Health Service, Agency for Toxic Substances and Disease Registry, 2006. Toxicological Profile for Hydrogen Sulfide, Atlanta, GA.

USEPA. 1998. Air Pollution Operating Permit Update. Key Features and Benefits. US Environmental Protection Agency, Office of Air Quality Planning and Standards, Research Triangle Park, N.C. EPA/451/K-98/002. February.

USEPA. 1987. National Primary Drinking Water Regulations; Synthetic Organic Chemicals; Monitoring for Unregulated Contaminants, Federal Register, Vol. 52, No. 130, July 7, 1987 Washington D.C.

USEPA (US Environmental Protection Agency). 2013. GAC Isotherms. Drinking Water Treatability Database. http://iaspub.epa.gov/tdb/pages/treatment/treatmentOverview.do?treatmentProcessId=-979193564.

USEPA. 2013a. Pollutants and Sources. Technology Transfer Network – Air Toxics Web Site. US Environmental Protection Agency. Office of Air and Radiation. http://www.epa.gov/ttn/atw/pollsour.html (accessed September 2013).

USEPA. 2013b. The Clean Air Act Amendments of 1990 List of Hazardous Air Pollutants. Technology Transfer Network – Air Toxics Web Site. US Environmental Protection Agency. Office of Air and Radiation, http://www.epa.gov/ttn/atw/orig189.html.

USEPA. 2000. Wastewater technology fact sheet. Ammonia stripping. U.S. Environmental Protection Agency, Washington, DC (EPA 832/F-00/019).

WHO (World Health Organization). 2003. Hydrogen Sulfide: Human Health Aspects. Concise International Chemical Assessment Document 53. Geneva.

Zogorski, J.S.; Carter, J.M.; Ivahnenko, T.; Lapham, W.W.; Moran, M.J.; Rowe, B.L.; Squillace, P.J.; and Toccalino, P.L. 2006. Volatile Organic Compounds in the Nation's Ground Water and Drinking Water Supply Wells, Circular 1292, United States Geological Survey, Reston, VA.

Chapter 5

Installation, Operation, and Maintenance

This chapter covers practical aspects of installation and operation and maintenance (O&M) of aeration facilities, with a focus on gas removal (air stripping) facilities. These facilities typically include major equipment besides the air stripping system (e.g., packed tower, tray, or diffused bubble aerators, all with blowers) such as source water pumping, finished water pumping, electrical systems, instrumentation and control (including telemetry) systems, and, if required, off-gas handling systems (e.g., gas phase GAC contactors, catalytic oxidizers, etc.) and chemical feed and storage systems. Topics covered provide owners and operators with guidance on key issues to assist them in attaining successful aeration projects. Two case studies illustrate problems encountered in implementing aeration (air stripping) projects.

INSTALLATION

Installation includes construction, commissioning, and startup of the aeration systems and associated facilities. Owner/operators generally assume construction administration and observation oversight to ensure that contractors observe design criteria faithfully and rigorously. Aeration facilities can include several subsystems that must function together seamlessly for the aeration system to perform as designed. Some construction contracts require a "system integrator", typically a control system technician or engineer who acts as a clearinghouse for all mechanical, process, electrical system functions, and ensures they rigorously relate to all physical/analog and electronic controls during the startup and commissioning phase. In addition, some owners have benefited from including a licensed operator, experienced with the specific aeration unit process, during startup. Detailed discussion of commissioning and startup planning is beyond the scope of this manual and is covered elsewhere (Kawamura 2000, HDR 2010, USEPA 1991).

Owners and operators should develop an implementation plan for installation of a new aeration facility or of adding an aeration process to an existing facility. Key features of the plan should include

- Ensuring that construction meets the design intent with respect to regulatory (SDWA/State rules), design criteria for treatment goals and occupational safety (OSHA), and equipment accessibility. Some key aeration facility characteristics to check include
    - Material and height of packing in aeration towers meet specifications and space above the packing is adequate to distribute water evenly over the packing material.
    - Bubble diffusers are at the specified depth to provide proper air-water exchange.
    - Electrical safety procedures should be emphasized for treatment processes with live electrical loads.
    - Equipment placement allows performance of operations and maintenance (O&M) activities such operating valves, servicing blowers, and periodically removing major equipment from buildings.
- Implementing comprehensive commissioning program to demonstrate that system functions and the components that support such functions are integrated and are operating properly. A stepwise approach from small sub-components to larger unit processes
- Observing system startup, including control strategies, and providing operators with system-specific training to facilitate unaided operation. Performance testing should demonstrate that design conditions can be achieved, including any startup contaminant spiking that may occur.

OPERATIONS AND MAINTENANCE

Operations and maintenance (O&M) of aeration treatment systems have generally become a mature practice element of drinking water treatment. The following discussion provide a broad guide to focus individually on specific operating locations.

Operational Guidance

Operation of aeration facilities may include the following:

- Operators to inspect, monitor, and adjust aeration equipment and blowers as well as supporting systems such as chemical feed for adjusting pH and/or disinfection, pumping, instrumentation and control, and waste gas treatment.
- Chemicals for pH adjustment before or after aeration and for disinfection and
- Electrical energy to power blowers, pumps, and instrumentation and control systems.
- Gas phase GAC or natural gas for catalytic oxidizers for facilities with off-gas treatment.
- Water and, possibly air, sampling for compliance monitoring and process control.

Table 5-1 Operational process parameters—monitoring for aeration treatment processes

Technology	Water flowrate	Water pressure	Air flowrate	Air pressure	Water pH	Air temperature	Hardness	Other
Packed Tower	Weekly	Not needed	Weekly	Weekly	Daily/ Weekly	Weekly	Weekly/ monthly	
Diffused Bubble	Weekly	Not needed	Weekly	Weekly	Daily/ Weekly	Weekly	Weekly/ monthly	
Tray	Weekly	Not needed	Weekly	Weekly	Daily/ Weekly	Weekly	Weekly/ monthly	
Spray	Weekly	Daily	Not needed	Not needed	Daily/ Weekly	Weekly	Weekly/ monthly	High-hardness process waters may foul (encrust) spray nozzles.
Membrane	Weekly	Daily	Weekly	Weekly	Daily/ Weekly	Weekly	Weekly/ monthly	Water-side pressure drop; water temperature
Venturi	Weekly	Daily	Weekly	Weekly	Daily/ Weekly	Weekly	Weekly/ monthly	

Note: In undertaking any installation, operation, and/or maintenance activity related to live electrical loads, stored mechanical or hydraulic energy, and/or hazardous chemicals, review and observe federal OSHA codes, guidelines, and best practices, in addition to any state or province-specific health and safety regulations for the industrial workplace.

Operation of aeration systems is typically straightforward and allows use of automated instrumentation and control systems. Providing remote monitoring may reduce the required amount of on-site operator attention, which is important for unattended treatment facilities.

Table 5-1 provides a general template for the routine monitoring of process parameters related to process water and finished water quality for several aeration technologies used in drinking water treatment. Operators should use this template as a guide for developing a process monitoring program for their individual facility consistent with their specific operating permit, aeration technology, and process control strategy.

Maintenance Guidance

Maintaining aeration facilities is a key aspect of achieving operational goals and regulatory compliance relevant to drinking water regulations. This includes routine scheduled maintenance such as air filter replacement and lubrication of motor bearings, periodic checking on physical condition of components such as tower structures, packing material, coatings, and anchor lugs, and troubleshooting declines in process performance due to biological fouling and mineral scaling and material failures.

Appendix B provides typical maintenance and troubleshooting guidance relative to the upkeep of these physical system components for several aeration technologies (packed towers, tray, diffused bubble, spray lateral and nozzle, and venturi aeration) and blowers or compressors and associated ductwork.

CASE STUDIES

Case Study No. 1—Aeration Equipment Materials Selection for an Aggressive Raw Water

Background. The Wortendyke Water Treatment Plant (WWTP) is a groundwater treatment facility, owned and operated by the Ridgewood Water Department, located in Bergen County, New Jersey. The plant receives flow from several sources, locally named Wells No. 2, 3, 6, and 7. The combined maximum discharge of these wells is 1,110 gpm, or roughly 1.6 MGD. The wells are drilled into the same rock formation and exhibit varying concentrations of the same list of raw water contaminants, all volatile organics regulated under the SDWA, and under NJ Administrative Code (the state-level regulatory document with NJ-delegated primacy for state enforcement of the SDWA). These contaminants are listed with their MCLs as enforced in New Jersey:

- PCE (tetrachloroethylene, MCL = 1 ppb)
- TCE (trichloroethylene, MCL = 5 ppb)
- MTBE (methyl tertiary butyl ether, MCL = 70 ppb)
- 1,1,1 TCA (1,1,1-Trichloroethane, MCL = 30 ppb)
- cis-1,2 DCE (1,2-Dichloroethene, MCL = 2 ppb)

The raw water mixture results from wells being drilled into a surficial rock aquifer in an historically industrial sector of the municipal service area. The original WWTP (construction date unknown) was simply a disinfection facility for the cited wells; however, in 1991 the facility was overhauled and improved to include a packed tower aerator, designed to remove the known regulated volatiles at the time, with PCE (at a raw water concentration of 15 ppb) as the basis-of-design contaminant. Figure 5-1 shows the combined raw water VOC data from April 1991 to July 2003.

Analysis: Air intake. At the time of design execution (1991), it was understood that the various sources exhibited multiple contaminant constituents and that further regulatory oversight of VOC's would be probable in future rulemakings. As a result, based on the typical methods published in industry texts (Kawamura 2002, MWH 2005), the stripping tower was installed with extra capacity when compared with the 15 pbb influent PCE concentration used as basis-of-design. The tower was installed at an overall height of 34 ft and a diameter of 7–8 in., treating the maximum flow rate cited above of 1,110 gpm, and an A:W ratio of 75:1.

In the early 1990s, Ridgewood Water began to test for and detect a varying concentration of methyl tertiary butyl ether (MTBE) in the blended raw water stream at WWTP. In approximately 1998, as an understanding of MTBE became more rigorous from a research standpoint and its response to aeration was better documented, Ridgewood Water took the proactive step of increasing the operational A:W ratio to 200:1. This stepped down the plant effluent MTBE concentration from approximately 16 ppb to 8 ppb. This step also required a significantly larger blower apparatus and increased the air intake velocity to the plant's blower room, which drew ambient atmospheric air through a filtered intake louver, of dimensions 4 × 8 ft.

The proposed increase in A:W ratio allowed a repeatable, predictable lowering in the effluent concentration of MTBE without the need to re-construct the stripping tower. However, it produced the negative countereffect of a higher-than-desired transfer velocity across the plant's air intake louver. At an A:W ratio of 75:1, the original design process air flow rate was approximately 11,000 cfm, resulting in an intake velocity at the original 4 × 8 ft louver of 5.79 fps. At intake velocities in excess of 10–12 fps, louvers often begin to exhibit water intake during rain (unless specifically designed to exclude water). Another operational difficulty is the matting of leaves and organic plant detritus to the intake (exterior) side of the louver during periods of blower activity—the air intake draws leaf

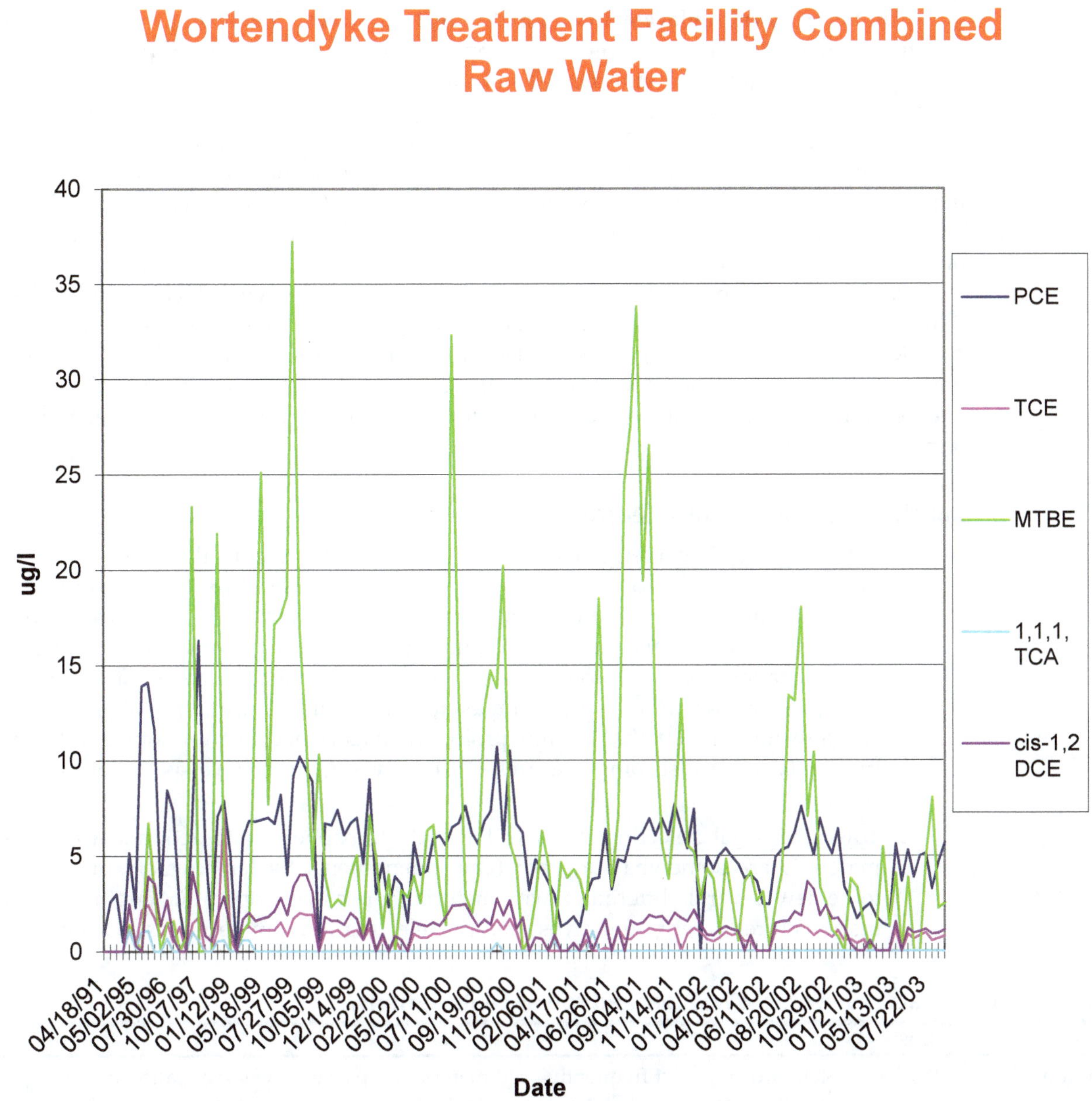

Source: Ridgewood Water Company

Figure 5-1 Wortendyke raw water quality 1991–2003

matter directly onto the screened louver, masking off the available louver area further, starving the blower of its source of air.

By increasing the A:W ratio to 200:1, the intake velocity was increased to 15.44 fps. As an unattended treatment station, it is impossible to monitor the louver status on more than a daily basis. As a result, Ridgewood Water decided in 2009 to double the louver size to a nominal surface area of 8 × 8 ft, resulting in an intake velocity of 7.72 fps. This solution allowed Ridgewood Water to continue operating at a higher A:W ratio and continue substantial removal of MTBE without increasing tower size, without the need to increase operational expenses via frequent manned visits to the station to check on louver masking.

Analysis: Tower replacement and material alternatives. Concurrent with the louver retrofit, Ridgewood Water replaced the stripping tower due to pitting of the aluminum column surface, resulting in pinhole leaks of process water through the tower. While not an initial water quality concern (air leakage was negligible), the pinholes were considered to be a forerunner to a potential future problem, if pitting were to become more frequent, or if the initial pinholes were to increase in diameter. Ridgewood Water reviewed several materials options on the basis of cost, constructability, and O&M considerations, with the object of replacing the existing tower in-kind dimensionally and functionally. A summary of the tower materials review is presented in Table 5-2.

ASCE—American Society of Civil Engineers, PVC—polyvinyl chloride, RCP—reinforced concrete pipe, VOCs—volatile active compounds, WTP—water treatment plant

Ridgewood Water selected the stainless steel option, based on its predicted chemical inertness, its local track record for positive performance in potable treatment applications, and its similarity in terms of structural characteristics and materials behavior over a range of potential weather conditions.

Conclusions and Lessons Learned

1. Air Handling Considerations: Process air handling is critical to the operation of an aeration unit process, and as such, all of the governing design parameters related to the air mover (either compressor, blower, or fan), as well as its accessory components (i.e., intake louver to plenum) must be taken into consideration when a modification is made to one of the process performance characteristics (in this example, air flow rate). To prevent masking of the air intake and possible drafting of rainwater, ensure that design project managers account for intake velocity, proximity to landscaping features, and louver maintenance as major design criteria.
2. Tower Material Selection: Raw water and atmospheric interaction with tower materials are somewhat unpredictable phenomena. System owners may wish to review several different tower material types with operator, engineer, and manufacturer's representatives before selecting one material as basis-of-bid

Table 5-2 Materials review for WWTP packed tower replacement

Material	Cost	Notes
Aluminum	$150,000	Standard and most frequently used material for packed tower fabrication in the northeastern United States. However, this material already exhibited a pitting tendency after prolonged exposure to the WTP's raw water
Stainless Steel	$225,000	Generally, less chemically reactive, but also more expensive with respect to raw materials, typically by 50% when benchmarked against aluminum
Concrete	$200,000	Somewhat novel design concept; as such, cost is conceptual rather than being based on known installations or even contractor estimates. Would require tower fabrication via custom casting, or by tapping air, water, and personnel access ports on an existing RCP pipe/tunnel piece, then fitting it with a roof finial and accessories for process water distribution, and process air demisting/venting. Essentially an estimated retrofit of standard metallic (here, assume 316 SS) mechanical internals to two (2) 20-ft lengths of 8-ft diameter RCP
PVC	$100,000	Infrequently used in the northeast United States. Owner expressed some concerns regarding structural degradation of the PVC by prolonged exposure to process water VOCs
Fiberglass	$150,000	Infrequently used in the northeast United States. Owner expressed concern regarding structural integrity given ASCE wind load rating for hurricane-prone regions (110–120 mph)

for a new packed tower application. Stakeholders should use benchmarking of materials interaction with similar raw water qualities in local applications, and place material integrity ahead of capital cost as a design criterion to mitigate the possibility of tower material loss, or potential process failure or structural failure.

Case Study No. 2—Limitations of Air Stripping Technology

Background. The Dobbs Avenue Station (DAS) is a groundwater treatment plant, constructed in 1991 in response to the detection of VOC's at an existing source, known locally as Well No. 5. The well operates at a capacity of 460 gpm. The treatment process was designed as packed tower aeration to remove tetrachloroethylene (PCE) and benzene.

As a result of a leaking fuel tank at a nearby gas station, MTBE was detected in the early 2000s. In 2003 the blower size was increased in an effort to increase A:W ratio and removal efficiency. The MTBE plume was sampled at ambient concentrations of 300–400 ppb, with spikes in the range of 500–1,000 ppb. With a best removal efficiency of 88 percent, the worst-case plant effluent of 120 ppb would exceed the NJ MCL of 70 ppb, as the plume drifted into Well No. 5's cone of depression.

Additionally, neighbors had lodged several complaints regarding airborne odors in the vicinity of the plant, with varying descriptions of "musty" and "rotten vegetables." The inspectors associated with NJ-state supervised cleanup of the MTBE spill characterized the odors as "MTBE-like." NJAW commissioned an air quality study to identify chemical composition of the waste air stream at DAS, and to assist in the design of any potential modifications to its treatment processes. The study results indicated an MTBE concentration of 225 ppb in the existing tower air discharge. Concentrations at the plant property boundaries were below the field equipment detection limit of 25 parts per trillion (ppt). The air quality study additionally identified the molecule 2-isopropyl 3-methoxy pyrazine (IPMP), an odor-causing compound also detected in the plant's raw water. Table 5-3 shows the concentrations of raw water contaminants.

From the air quality study, note the following deductions:

> *The existing air stripper had been retrofitted in 2003 with a larger blower [increased from 5 HP to 60 HP] due to increasing concentrations of MTBE in the groundwater. The size of the air stripper itself was not increased due to physical constraints in the size of the building. This resulted in a significant increase in air volume and velocity. During operation, significant water discharges were observed from the stack and from the clear well overflow. Given the large increase in air flow and the poor condition of the mist eliminator, we may conclude that the reported odors in the immediate vicinity of the site were due to concentrations of MTBE and/or IPMP exceeding odor thresholds in the water phase carried by water droplets discharged through the stack. It is not likely that odors could be detected from vapor phase stack discharges. The odor threshold for MTBE in air is reported to be in the 50 to 200 ug/L (ppb) range which is well above the <25.0 ppt ambient air levels determined during this study. The MTBE odor threshold in water, however,*

Table 5-3 Raw water contaminants

Contaminant	Raw Water Concentrations	NJ MCLs (μg/L)
MTBE	300–1,000 μg/L	70
PCE	1.6–3.1 μg/L	1
Benzene	0.6–3.2 μg/L	1
IPMP	3.4–3.7 ng/L	N/A

IPMP—2-isopropyl 3-methoxy pyrazine, MTBE— methyl tertiary butyl ether, PCE—tetrachloroethylene
Source: AWWSC 2007

> *is reported to be 20 to 40 ppb which is equivalent to the highest MTBE level of 40 ppb found in the finished water discharged from the air stripper. Similarly, the IPMP odor threshold in water is reported to be 2 ng/L (ppt), which is the approximate concentration of IPMP found in the finished water. The odor threshold for IPMP in air was not found in the literature... Further evidence that the odor is water based was provided by odor complaints received by New Jersey American at water tap location(s) remote from the site* (ENSR 2007).

Analysis: Taste and odor compounds as limiters for aeration technology. Analysis of the volatility of IPMP was beyond the scope of NJAW's investigation, and a Henry's law constant or similar data currently exists nowhere in the academic literature. However, anecdotal evidence from the Dobbs Avenue review indicates that IPMP showed no practical tendency to volatilize, even at the relatively high A:W ratio used to strip MTBE. As such— and accounting for customer complaints related to odor in the finished water— NJAW decided to include a second major treatment step in the process train at Dobbs Avenue Station; namely IPMP adsorption via GAC contactors. IPMP was identified as a taste and odor (T/O) contaminant at a single existing location in NJ and was removed to non-detect levels via an Empty Bed Contact Time (EBCT) of 15 minutes. NJAW utilized this information as its design criteria for Dobbs Avenue.

Analysis: Stripping a highly soluble contaminant in a residential neighborhood. At an ambient concentration of 300–400 ppb, NJAW might have continued to rely upon its existing packed tower to remove MTBE. However, with the potential increase to a range of 500–1,000 ppb, NJAW made the decision to increase its treatment efficiencies to ensure complete compliance with NJ's MCL of 70 ppb. With the added concern of finished water odor, NJAW set the process design criteria at a maximum 20 ppb MTBE in finished water, with a raw water concentration of 1,000 ppb. These worst-case concentrations resulted in a tower size that significantly exceeded the municipal structural height ordinance of 35 ft. With approximately 40 ft of minimum required height for the proposed tower packing material, NJAW faced a minimum practical overall tower height of 50′ (accounting for distribution mechanicals and support materials at the base). Anticipating a rejection for height variance from the municipal zoning board, NJAW approach the problem with a novel solution: split the packing height into two (2) towers constructed in-series (see Figure 5-2). Each tower would be constructed at a height of 30 ft (allowing installation under a code-compliant 35-ft roofline).

Raw water would flow into the first tower, with the first tower collection cellar acting as a clearwell for a booster pump. The booster pump would send process water to the second tower for final discharge to a below-grade plant clearwell. A single blower would be shared between the two towers, drawing air by suction through the second tower, then discharging it to the first tower. This innovation required an additional pumping step to accomplish the required MTBE removal. However, it allowed NJAW to move ahead with station operation in a residential neighborhood whose quite common zoning ordinances might otherwise have prohibited an increase in stripping tower height.

Conclusions and Lessons Learned

1. Limits of Aeration Technology: A commonplace assumption in the operations community is that aeration is a catch-all technology for elimination of waterborne odors. However, some T/O compounds present the difficult challenge of low odor detection threshold to human consumers and little or no tendency to volatilize. In addressing T/O compounds, operators may wish to review the literature for the responsiveness of different chemical species to aeration—or at a minimum conduct a bench-scale removal study—before undertaking full-scale construction of aeration technology as a T/O removal unit process. Liquid-phase GAC is a well-proven alternative unit process in situations where aeration technology is not effective in T/O compound removal.

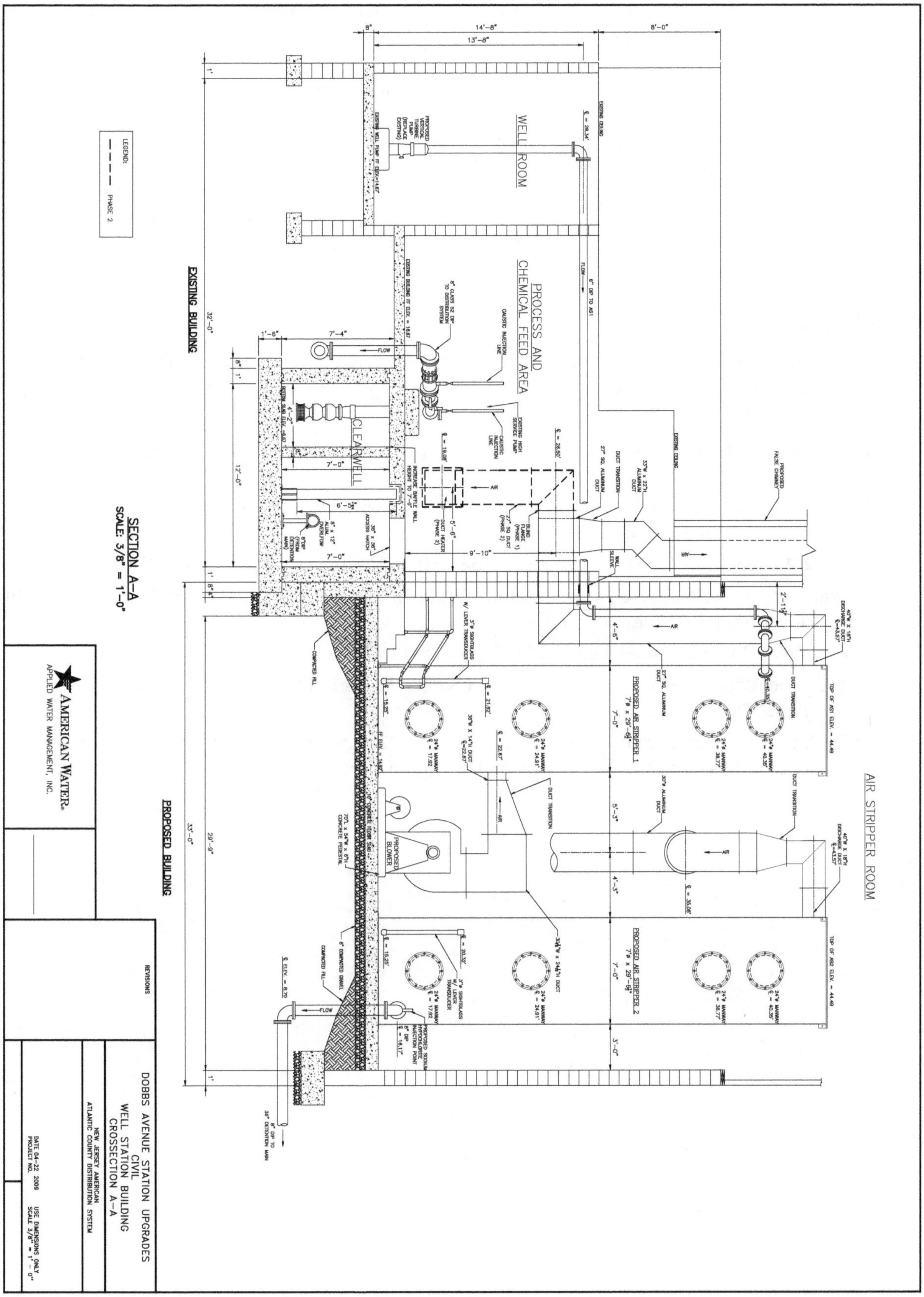

Source: AWM 2012

Figure 5-2 Dobbs Avenue treatment process

2. Packed Tower Construction vs Height Limitations: In many municipalities, structural height and building appearance are highly regulated, especially in areas zoned residential. Operators must weigh these limitations against the value of individual groundwater sources, when confronted with evidence of contamination by VOC's. If a source is of high value, the operator may consider staging the stripping process in series. This alternative trades off the operational expense of a second pumping stage for the value of maintaining water quality compliance in locations with no tolerance for structural height zoning ordinance variances. Further, operators should note that process air can be moved countercurrently through the series of packed towers via a single blower.

REFERENCES

AWM (Applied Water Management). 2012. Dobbs Avenue Station Treatment Process Upgrades.

AWWSC (American Waterworks Service Company). 2007. Dobbs Avenue Station, "Design Concept," August 2007.

ENSR (AECOM). 2007. Dobbs Avenue Air Stripper Source and Ambient Sampling Results, May 2007.

Havens & Emerson. 1990. Contract No. 1 Plans: Twinney, East Ridgewood, and Wortendyke Treatment Facilities, October 1990.

HDR. 2010. Operation & Maintenance Manual, City of Redmond, Washington Wells No. 1 and 2.

Kawamura, S. 2000 (2nd ed.). Integrated Design and Operation of Water Treatment Facilities. New York: John Wiley & Sons.

MWH. 2005 (2nd ed.). *Water Treatment Principles and Design.*

Ridgewood Water Department. n.d. Water Quality Historic Data Record, April 18, 1991 through July 22, 2003.

USEPA (US Environmental Protection Agency). 1991. Air Stripping of Aqueous Solutions. EPA Office of Research and Development. Cincinnati, OH.

Appendix A

Dose-Response Values

Table A-1 Prioritized chronic dose-response values (5/21/2012)

Revisions since 04/27/10 are shown in red.
CAS NO. = Chemical Abstracts Services number for the compound.
HAP NO. = Position of the compound on the HAP list in the Clean Air Act (112[b][2]). "999" denotes substances under consideration for listing.
Sources: IRIS = Integrated Risk Information System; ATSDR = US Agency for Toxic Substances and Disease Registry; D-ATSDR = draft ATSDR; CA = California EPA; P-CAL = Proposed CAL; HEAST = EPA Health Effects Assessment Tables

IARC WOE = weight of evidence for carcinogenicity in humans (1 - carcinogenic; 2A - probably carcinogenic; 2B - possibly carcinogenic; 3 - not classifiable; 4 - probably not carcinogenic).

EPA WOE (2005 Guidelines) = weight of evidence for carcinogenicity under 2005 EPA cancer guidelines:
CH - carcinogenic to humans; LH - likely to be carcinogenic;
SE - suggestive evidence of carcinogenic potential;
InI - inadequate information to assess carcinogenic potential;
NH - not likely to be carcinogenic).
EPA MOA (2005 Guidelines) = mode of action for carcinogenicity:
M* - mutagenic and early life data lacking; age-dependent adjustment factors should be applied when assessing risk for ages younger than 16 years per 2005 Supplemental Guidance.
[See Table4A-1 notes.]

EPA WOE (1986 Guidelines) = weight of evidence for carcinogenicity under the 1986 EPA cancer guidelines:
A - human carcinogen;
B1 - probable carcinogen, limited human evidence;
B2 - probable carcinogen, sufficient evidence in animals;
C - possible human carcinogen; D - not classifiable
E - evidence of noncarcinogenicity.

				CHRONIC INHALATION						CHRONIC ORAL					
				NONCANCER		CANCER				NONCANCER		CANCER			
CHEMICAL NAME	CAS NO.	HAP NO.	IARC WOE	mg/m^3	SOURCE	EPA WOE	$1/(ug/m^3)$	MOA	SOURCE	mg/kg/d	SOURCE	EPA WOE	1/(mg/kg/d)	MOA	SOURCE
Acetaldehyde	75-07-0	1	2B	0.009	IRIS	B2	2.2E-06		IRIS						
Acetamide	60-35-5	2	2B				0.00002		CAL						
Acetonitrile	75-05-8	3		0.06	IRIS	InI									
Acetophenone	98-86-2	4				D									
Acrolein	107-02-8	6	3	0.00002	IRIS	InI									
Acrylamide	79-06-1	7	2A	0.006	IRIS	LH	0.0001	M*	IRIS						
Acrylic acid	79-10-7	8	3	0.001	IRIS										
Acrylonitrile	107-13-1	9	2B	0.002	IRIS	B1	0.000068		IRIS						
Allyl chloride	107-05-1	10	3	0.001	IRIS	C	0.000006		CAL						
Aniline	62-53-3	12	3	0.001	IRIS	B2	1.6E-06		CAL						
Antimony compounds	7440-36-0	173													
Antimony pentoxide	1314-60-9	173													
Antimony potassium tartrate	304-61-0	173													
Antimony tetroxide	1332-81-6	173													
Antimony trioxide	1309-64-4	173	2B	0.0002	IRIS										
Arsenic compounds	7440-38-2	174	1	0.000015	CAL	A	0.0043		IRIS						
Arsenic pentoxide	1303-28-2	174													
Arsine	7784-42-1	174		0.00005	IRIS										
Benzene	71-43-2	15	1	0.03	IRIS	CH	7.8E-06		IRIS						
Benzidine	92-87-5	16	1	0.01	P-CAL	A	0.067	M*	IRIS						
Benzotrichloride	98-07-7	17	2B			B2									

Revisions since 04/27/10 are shown in red.
CAS NO. = Chemical Abstracts Services number for the compound.
HAP NO. = Position of the compound on the HAP list in the Clean Air Act (112[b][2]). "999" denotes substances under consideration for listing.
Sources: IRIS = Integrated Risk Information System; ATSDR = US Agency for Toxic Substances and Disease Registry; D-ATSDR = draft ATSDR; CA = California EPA; P-CAL = Proposed CAL; HEAST = EPA Health Effects Assessment Tables

IARC WOE = weight of evidence for carcinogenicity in humans (1 - carcinogenic; 2A - probably carcinogenic; 2B - possibly carcinogenic; 3 - not classifiable; 4 - probably not carcinogenic).

EPA WOE (2005 Guidelines) = weight of evidence for carcinogenicity under 2005 EPA cancer guidelines:
CH - carcinogenic to humans; LH - likely to be carcinogenic;
SE - suggestive evidence of carcinogenic potential;
InI - inadequate information to assess carcinogenic potential;
NH - not likely to be carcinogenic).
EPA MOA (2005 Guidelines) = mode of action for carcinogenicity:
M* - mutagenic and early life data lacking; age-dependent adjustment factors should be applied when assessing risk for ages younger than 16 years per 2005 Supplemental Guidance.
[See Table4A-1 notes.]

EPA WOE (1986 Guidelines) = weight of evidence for carcinogenicity under the 1986 EPA cancer guidelines:
A - human carcinogen;
B1 - probable carcinogen, limited human evidence;
B2 - probable carcinogen, sufficient evidence in animals;
C - possible human carcinogen; D - not classifiable
E - evidence of noncarcinogenicity.

				CHRONIC INHALATION						CHRONIC ORAL					
				NONCANCER		CANCER				NONCANCER		CANCER			
CHEMICAL NAME	CAS NO.	HAP NO.	IARC WOE	mg/m^3	SOURCE	EPA WOE	1/(ug/m^3)	MOA	SOURCE	mg/kg/d	SOURCE	EPA WOE	1/(mg/kg/d)	MOA	SOURCE
Benzyl chloride	100-44-7	18	2B			B2	0.000049		CAL						
Beryllium compounds	7440-41-7	175	1	0.00002	IRIS	LH	0.0024		IRIS						
Beryllium oxide	1304-56-9	175													
Biphenyl	92-52-4	19				D									
Bis(2-ethylhexyl)phthalate	117-81-7	20	2B	0.01	P-CAL	B2	2.4E-06		CAL						
Bis(chloromethyl)ether	542-88-1	21	1			A	0.062		IRIS						
Bromoform	75-25-2	22	3			B2	1.1E-06		IRIS						
1,3-Butadiene	106-99-0	23	1	0.002	IRIS	CH	0.00003		IRIS						
Cadmium compounds	7440-43-9	176	1	0.00001	D-ATSDR	B1	0.0018		IRIS	0.0005	IRIS	B1			
Captan	133-06-2	26	3			B2									
Carbaryl	63-25-2	27													
Carbon disulfide	75-15-0	28		0.7	IRIS										
Carbon tetrachloride	56-23-5	29	2B	0.1	IRIS	LH	0.000006		IRIS						
Chloramben	133-90-4	32													
Chlordane	57-74-9	33	2B	0.0007	IRIS	LH	0.0001		IRIS	0.0005	IRIS	LH	0.35		IRIS
Chlorine	7782-50-5	34		0.00015	*ATSDR*										
Chloroacetic acid	79-11-8	35													
2-Chloroacetophenone	532-27-4	36		0.00003	IRIS										
Chlorobenzene	108-90-7	37		1	CAL	D									
Chlorobenzilate	510-15-6	38	3			B2	0.000078		HEAST						
Chloroform	67-66-3	39	2B	0.098	ATSDR	LH									

Revisions since 04/27/10 are shown in red.
CAS NO. = Chemical Abstracts Services number for the compound.
HAP NO. = Position of the compound on the HAP list in the Clean Air Act (112[b][2]). "999" denotes substances under consideration for listing.
Sources: IRIS = Integrated Risk Information System; ATSDR = US Agency for Toxic Substances and Disease Registry; D-ATSDR = draft ATSDR; CA = California EPA; P-CAL = Proposed CAL; HEAST = EPA Health Effects Assessment Tables

IARC WOE = weight of evidence for carcinogenicity in humans (1 - carcinogenic; 2A - probably carcinogenic; 2B - possibly carcinogenic; 3 - not classifiable; 4 - probably not carcinogenic).

EPA WOE (2005 Guidelines) = weight of evidence for carcinogenicity under 2005 EPA cancer guidelines: CH - carcinogenic to humans; LH - likely to be carcinogenic; SE - suggestive evidence of carcinogenic potential; InI - inadequate information to assess carcinogenic potential; NH - not likely to be carcinogenic).
EPA MOA (2005 Guidelines) = mode of action for carcinogenicity: M* - mutagenic and early life data lacking; age-dependent adjustment factors should be applied when assessing risk for ages younger than 16 years per 2005 Supplemental Guidance. *[See Table4A-1 notes.]*

EPA WOE (1986 Guidelines) = weight of evidence for carcinogenicity under the 1986 EPA cancer guidelines:
A - human carcinogen;
B1 - probable carcinogen, limited human evidence;
B2 - probable carcinogen, sufficient evidence in animals;
C - possible human carcinogen; D - not classifiable
E - evidence of noncarcinogenicity.

				CHRONIC INHALATION						CHRONIC ORAL					
				NONCANCER		CANCER				NONCANCER		CANCER			
CHEMICAL NAME	CAS NO.	HAP NO.	IARC WOE	mg/m^3	SOURCE	EPA WOE	1/(ug/m^3)	MOA	SOURCE	mg/kg/d	SOURCE	EPA WOE	1/(mg/kg/d)	MOA	SOURCE
Chloroprene	126-99-8	41	2B	0.02	IRIS	LH	0.0003	M*	IRIS						
Chromium (III) compounds	16065-83-1	177	3			InI									
Chromium (VI) compounds	18540-29-9	177	1	0.0001	IRIS	CH	0.012		IRIS						
Chromium (VI) trioxide, chromic acid mist	11115-74-5	177	1	0.000008	IRIS										
Cobalt compounds	7440-48-4	178	2B	0.0001	ATSDR										
Coke Oven Emissions	8007-45-2	179				A	0.00062		IRIS						
m-Cresol	108-39-4	44				C									
o-Cresol	95-48-7	43				C									
p-Cresol	106-44-5	45				C									
Cresols (mixed)	1319-77-3	42		0.6	CAL	C									
Cumene	98-82-8	46	2B	0.4	IRIS	InI									
Cyanazine	21725-46-2	180				C									
Cyanide compounds	57-12-5	180				D									
Acetone cyanohydrin	75-86-5	180		0.01	HEAST										
Calcium cyanide	592-01-8	180													
Copper cyanide	544-92-3	180													
Cyanogen	460-19-5	180		0.0008	IRIS										
Cyanogen bromide	506-68-3	180													
Cyanogen chloride	506-77-4	180													
Ethylene cyanohydrin	109-78-4	180													
Hydrogen cyanide	74-90-8	180		0.0008	IRIS										

Revisions since 04/27/10 are shown in red.
CAS NO. = Chemical Abstracts Services number for the compound.
HAP NO. = Position of the compound on the HAP list in the Clean Air Act (112[b][2]). "999" denotes substances under consideration for listing.
Sources: IRIS = Integrated Risk Information System; ATSDR = US Agency for Toxic Substances and Disease Registry; D-ATSDR = draft ATSDR; CA = California EPA; P-CAL = Proposed CAL; HEAST = EPA Health Effects Assessment Tables

IARC WOE = weight of evidence for carcinogenicity in humans (1 - carcinogenic; 2A - probably carcinogenic; 2B - possibly carcinogenic; 3 - not classifiable; 4 - probably not carcinogenic).

EPA WOE (2005 Guidelines) = weight of evidence for carcinogenicity under 2005 EPA cancer guidelines: CH - carcinogenic to humans; LH - likely to be carcinogenic; SE - suggestive evidence of carcinogenic potential; InI - inadequate information to assess carcinogenic potential; NH - not likely to be carcinogenic).
EPA MOA (2005 Guidelines) = mode of action for carcinogenicity:
M* - mutagenic and early life data lacking; age-dependent adjustment factors should be applied when assessing risk for ages younger than 16 years per 2005 Supplemental Guidance. *[See Table4A-1 notes.]*

EPA WOE (1986 Guidelines) = weight of evidence for carcinogenicity under the 1986 EPA cancer guidelines:
A - human carcinogen;
B1 - probable carcinogen, limited human evidence;
B2 - probable carcinogen, sufficient evidence in animals;
C - possible human carcinogen; D - not classifiable
E - evidence of noncarcinogenicity.

				CHRONIC INHALATION						CHRONIC ORAL					
				NONCANCER		CANCER				NONCANCER		CANCER			
CHEMICAL NAME	CAS NO.	HAP NO.	IARC WOE	mg/m^3	SOURCE	EPA WOE	1/(ug/m^3)	MOA	SOURCE	mg/kg/d	SOURCE	EPA WOE	1/(mg/kg/d)	MOA	SOURCE
Potassium cyanide	151-50-8	180		*0.0008*	IRIS										
Potassium silver cyanide	506-61-6	180		*0.0008*	IRIS										
Silver cyanide	506-64-9	180													
Sodium cyanide	143-33-9	180		*0.0008*	IRIS										
Thiocyanic acid, 2-(benzothiazolylthio) methyl est	21564-17-0	180													
Zinc cyanide	557-21-1	180													
2,4-D, salts and esters	94-75-7	47													
DDE	72-55-9	48				B2						B2	0.34		IRIS
1,2-Dibromo-3-chloropropane	96-12-8	51	2B	0.0002	IRIS	B2	0.002		CAL						
Dibutylphthalate	84-74-2	52				D									
p-Dichlorobenzene	106-46-7	53	2B	0.8	IRIS	C	0.000011		CAL						
3,3'-Dichlorobenzidine	91-94-1	54	2B			B2	0.00034		CAL						
Dichloroethyl ether	111-44-4	55	3			B2	0.00033		IRIS						
1,3-Dichloropropene	542-75-6	56	2B	0.02	IRIS	LH	0.000004		IRIS						
Dichlorvos	62-73-7	57	2B	0.0005	IRIS	B2									
Diesel engine emissions	DIESEL EMIS.	999		0.005	IRIS	LH									
Diethanolamine	111-42-2	58		0.003	CAL										
3,3'-Dimethoxybenzidine	119-90-4	61	2B			B2									

Revisions since 04/27/10 are shown in red.
CAS NO. = Chemical Abstracts Services number for the compound.
HAP NO. = Position of the compound on the HAP list in the Clean Air Act (112[b][2]). "999" denotes substances under consideration for listing.
Sources: IRIS = Integrated Risk Information System; ATSDR = US Agency for Toxic Substances and Disease Registry; D-ATSDR = draft ATSDR; CA = California EPA; P-CAL = Proposed CAL; HEAST = EPA Health Effects Assessment Tables

IARC WOE = weight of evidence for carcinogenicity in humans (1 - carcinogenic; 2A - probably carcinogenic; 2B - possibly carcinogenic; 3 - not classifiable; 4 - probably not carcinogenic).

EPA WOE (2005 Guidelines) = weight of evidence for carcinogenicity under 2005 EPA cancer guidelines:
CH - carcinogenic to humans; LH - likely to be carcinogenic;
SE - suggestive evidence of carcinogenic potential;
InI - inadequate information to assess carcinogenic potential;
NH - not likely to be carcinogenic).
EPA MOA (2005 Guidelines) = mode of action for carcinogenicity:
M* - mutagenic and early life data lacking; age-dependent adjustment factors should be applied when assessing risk for ages younger than 16 years per 2005 Supplemental Guidance. *[See Table4A-1 notes.]*

EPA WOE (1986 Guidelines) = weight of evidence for carcinogenicity under the 1986 EPA cancer guidelines:
A - human carcinogen;
B1 - probable carcinogen, limited human evidence;
B2 - probable carcinogen, sufficient evidence in animals;
C - possible human carcinogen; D - not classifiable
E - evidence of noncarcinogenicity.

				CHRONIC INHALATION						CHRONIC ORAL					
				NONCANCER		CANCER				NONCANCER		CANCER			
CHEMICAL NAME	CAS NO.	HAP NO.	IARC WOE	mg/m^3	SOURCE	EPA WOE	1/(ug/m^3)	MOA	SOURCE	mg/kg/d	SOURCE	EPA WOE	1/(mg/kg/d)	MOA	SOURCE
p-Dimethylaminoazobenzene	60-11-7	62	2B				0.0013		CAL						
3,3'-Dimethylbenzidine	119-93-7	63				B2									
Dimethyl formamide	68-12-2	65	3	0.03	IRIS										
N,N-dimethylaniline	121-69-7	59	3												
1,1-Dimethylhydrazine	57-14-7	66	2B			B2									
2,4-dinitrophenol	51-28-5	70													
2,4-Dinitrotoluene	121-14-2	71	2B	0.007	P-CAL	B2	0.000089		CAL						
2,4/2,6-Dinitrotoluene (mixture)	25321-14-6	71	2B			B2									
1,4-Dioxane	123-91-1	72	2B	3.6	D-ATSDR	LH									
1,2-Diphenylhydrazine	122-66-7	73				B2	0.00022		IRIS						
Epichlorohydrin	106-89-8	74	2A	0.001	IRIS	B2	1.2E-06		IRIS						
1,2-Epoxybutane	106-88-7	75		0.02	IRIS										
Ethyl acrylate	140-88-5	76	2B			B2									
Ethyl benzene	100-41-4	77	2B	1	IRIS	D	2.5E-06		CAL						
Ethyl carbamate	51-79-6	78	2A				0.00029	M*	CAL						
Ethyl chloride	75-00-3	79	3	10	IRIS										
Ethylene dibromide	106-93-4	80	2A	0.009	IRIS	LH	0.0006		IRIS						
Ethylene dichloride	107-06-2	81	2B	2.4	ATSDR	B2	0.000026		IRIS						
Ethylene glycol	107-21-1	82		0.4	CAL										
Ethylene oxide	75-21-8	84	1	0.03	CAL	B1	0.000088		CAL						
Ethylene thiourea	96-45-7	85	3	0.003	P-CAL	B2	0.000013		CAL						

Revisions since 04/27/10 are shown in red.
CAS NO. = Chemical Abstracts Services number for the compound.
HAP NO. = Position of the compound on the HAP list in the Clean Air Act (112[b][2]). "999" denotes substances under consideration for listing.
Sources: IRIS = Integrated Risk Information System; ATSDR = US Agency for Toxic Substances and Disease Registry; D-ATSDR = draft ATSDR; CA = California EPA; P-CAL = Proposed CAL; HEAST = EPA Health Effects Assessment Tables

IARC WOE = weight of evidence for carcinogenicity in humans (1 - carcinogenic; 2A - probably carcinogenic; 2B - possibly carcinogenic; 3 - not classifiable; 4 - probably not carcinogenic).

EPA WOE (2005 Guidelines) = weight of evidence for carcinogenicity under 2005 EPA cancer guidelines:
CH - carcinogenic to humans; LH - likely to be carcinogenic;
SE - suggestive evidence of carcinogenic potential;
InI - inadequate information to assess carcinogenic potential;
NH - not likely to be carcinogenic).
EPA MOA (2005 Guidelines) = mode of action for carcinogenicity:
M* - mutagenic and early life data lacking; age-dependent adjustment factors should be applied when assessing risk for ages younger than 16 years per 2005 Supplemental Guidance.
[See Table4A-1 notes.]

EPA WOE (1986 Guidelines) = weight of evidence for carcinogenicity under the 1986 EPA cancer guidelines:
A - human carcinogen;
B1 - probable carcinogen, limited human evidence;
B2 - probable carcinogen, sufficient evidence in animals;
C - possible human carcinogen; D - not classifiable
E - evidence of noncarcinogenicity.

				CHRONIC INHALATION						CHRONIC ORAL					
				NONCANCER		CANCER				NONCANCER		CANCER			
CHEMICAL NAME	CAS NO.	HAP NO.	IARC WOE	mg/m^3	SOURCE	EPA WOE	1/(ug/m^3)	MOA	SOURCE	mg/kg/d	SOURCE	EPA WOE	1/(mg/kg/d)	MOA	SOURCE
Ethylidene dichloride (1,1-Dichloroethane)	75-34-3	86		0.5	HEAST	C	1.6E-06		CAL						
Formaldehyde	50-00-0	87	1	0.0098	ATSDR	B1	0.000013		IRIS						
Diethylene glycol monobutyl ether	112-34-5	181		0.02	HEAST										
Diethylene glycol monoethyl ether	111-90-0	181													
Ethylene glycol ethyl ether	110-80-5	181		0.2	IRIS										
Ethylene glycol ethyl ether acetate	111-15-9	181		0.3	CAL										
Ethylene glycol methyl ether	109-86-4	181		0.02	IRIS										
Ethylene glycol methyl ether acetate	110-49-6	181		0.09	CAL										
Heptachlor	76-44-8	88	2B			B2	0.0013		IRIS	0.0005	IRIS	B2	4.5		IRIS
Hexachlorobenzene	118-74-1	89	2B	0.003	P-CAL	B2	0.00046		IRIS	0.0008	IRIS	B2	1.6		IRIS
Hexachlorobutadiene	87-68-3	90	3	0.09	P-CAL	C	0.000022		IRIS						
Hexachlorocyclopentadiene	77-47-4	91		0.0002	IRIS	NH									
Hexachlorodibenzo-p-dioxin, mixture	19408-74-3	187				B2	1.3		IRIS			B2	6200		IRIS
Hexachloroethane	67-72-1	92	2B	*0.03*	*IRIS*	*LH*									
Hexamethylene-1,6-diisocyanate	822-06-0	93		0.00001	IRIS										

Revisions since 04/27/10 are shown in red.
CAS NO. = Chemical Abstracts Services number for the compound.
HAP NO. = Position of the compound on the HAP list in the Clean Air Act (112[b][2]). "999" denotes substances under consideration for listing.
Sources: IRIS = Integrated Risk Information System; ATSDR = US Agency for Toxic Substances and Disease Registry; D-ATSDR = draft ATSDR; CA = California EPA; P-CAL = Proposed CAL; HEAST = EPA Health Effects Assessment Tables

IARC WOE = weight of evidence for carcinogenicity in humans (1 - carcinogenic; 2A - probably carcinogenic; 2B - possibly carcinogenic; 3 - not classifiable; 4 - probably not carcinogenic).

EPA WOE (2005 Guidelines) = weight of evidence for carcinogenicity under 2005 EPA cancer guidelines:
CH - carcinogenic to humans; LH - likely to be carcinogenic;
SE - suggestive evidence of carcinogenic potential;
InI - inadequate information to assess carcinogenic potential;
NH - not likely to be carcinogenic).
EPA MOA (2005 Guidelines) = mode of action for carcinogenicity:
M* - mutagenic and early life data lacking; age-dependent adjustment factors should be applied when assessing risk for ages younger than 16 years per 2005 Supplemental Guidance. *[See Table4A-1 notes.]*

EPA WOE (1986 Guidelines) = weight of evidence for carcinogenicity under the 1986 EPA cancer guidelines:
A - human carcinogen;
B1 - probable carcinogen, limited human evidence;
B2 - probable carcinogen, sufficient evidence in animals;
C - possible human carcinogen; D - not classifiable
E - evidence of noncarcinogenicity.

				CHRONIC INHALATION						CHRONIC ORAL					
				NONCANCER		CANCER				NONCANCER		CANCER			
CHEMICAL NAME	CAS NO.	HAP NO.	IARC WOE	mg/m^3	SOURCE	EPA WOE	1/(ug/m^3)	MOA	SOURCE	mg/kg/d	SOURCE	EPA WOE	1/(mg/kg/d)	MOA	SOURCE
n-Hexane	110-54-3	95		0.7	IRIS	InI									
Hydrazine	302-01-2	96	2B	0.0002	CAL	B2	0.0049		IRIS						
Hydrochloric acid	7647-01-0	97	3	0.02	IRIS										
Hydrofluoric acid	7664-39-3	98		0.014	CAL										
Hydrogen sulfide	7783-06-4	999		0.002	IRIS	InI									
Hydroquinone	123-31-9	99													
Isophorone	78-59-1	100		2	CAL	C									
Lead compounds	7439-92-1	182	2B	0.00015	EPA OAQPS	B2						B2			
Tetraethyl lead	78-00-2	182								0.0000001	IRIS				
Lindane (gamma-HCH)	58-89-9	101	2B	0.0003	P-CAL	B2-C	0.00031		CAL	0.0003	IRIS	B2-C	1.1		CAL
alpha-Hexachlorocyclohexane (a-HCH)	319-84-6	101	2B	0.02	P-CAL	B2	0.0018		IRIS	0.008	D-ATSDR	B2	6.3		IRIS
beta-Hexachlorocyclohexane (b-HCH)	319-85-7	101	2B	0.002	P-CAL	C	0.00053		IRIS			C	1.8		IRIS
technical Hexachlorocyclohexane (HCH)	608-73-1	101	2B			B2	0.00051		IRIS			B2	1.8		IRIS
Maleic anhydride	108-31-6	102		0.0007	CAL										
Manganese compounds	7439-96-5	183		0.00005	IRIS	D									
Mercuric chloride	7487-94-7	184				C				0.0003	IRIS	C			
Mercury (elemental)	7439-97-6	184	3	0.0003	IRIS	D						D			
Methyl mercury	22967-92-6	184				C				0.0001	IRIS	C			

Revisions since 04/27/10 are shown in red.
CAS NO. = Chemical Abstracts Services number for the compound.
HAP NO. = Position of the compound on the HAP list in the Clean Air Act (112[b][2]). "999" denotes substances under consideration for listing.
Sources: IRIS = Integrated Risk Information System; ATSDR = US Agency for Toxic Substances and Disease Registry; D-ATSDR = draft ATSDR; CA = California EPA; P-CAL = Proposed CAL; HEAST = EPA Health Effects Assessment Tables

IARC WOE = weight of evidence for carcinogenicity in humans (1 - carcinogenic; 2A - probably carcinogenic; 2B - possibly carcinogenic; 3 - not classifiable; 4 - probably not carcinogenic).

EPA WOE (2005 Guidelines) = weight of evidence for carcinogenicity under 2005 EPA cancer guidelines:
CH - carcinogenic to humans; LH - likely to be carcinogenic;
SE - suggestive evidence of carcinogenic potential;
InI - inadequate information to assess carcinogenic potential;
NH - not likely to be carcinogenic).
EPA MOA (2005 Guidelines) = mode of action for carcinogenicity:
M* - mutagenic and early life data lacking; age-dependent adjustment factors should be applied when assessing risk for ages younger than 16 years per 2005 Supplemental Guidance.
[See Table4A-1 notes.]

EPA WOE (1986 Guidelines) = weight of evidence for carcinogenicity under the 1986 EPA cancer guidelines:
A - human carcinogen;
B1 - probable carcinogen, limited human evidence;
B2 - probable carcinogen, sufficient evidence in animals;
C - possible human carcinogen; D - not classifiable
E - evidence of noncarcinogenicity.

				CHRONIC INHALATION						CHRONIC ORAL					
				NONCANCER		CANCER				NONCANCER		CANCER			
CHEMICAL NAME	CAS NO.	HAP NO.	IARC WOE	mg/m^3	SOURCE	EPA WOE	1/(ug/m^3)	MOA	SOURCE	mg/kg/d	SOURCE	EPA WOE	1/(mg/kg/d)	MOA	SOURCE
Phenylmercuric acetate	62-38-4	184								0.00008	IRIS				
Methanol	67-56-1	103		4	CAL										
Methoxychlor	72-43-5	104	3			D				0.005	IRIS	D			
Methyl bromide	74-83-9	105		0.005	IRIS	D									
Methyl chloride	74-87-3	106	3	0.09	IRIS	InI									
Methyl chloroform (1,1,1-Trichloroethane)	71-55-6	107		5	IRIS	InI									
Methyl isobutyl ketone	108-10-1	111	2B	3	IRIS	InI									
Methyl isocyanate	624-83-9	112		0.001	CAL										
Methyl methacrylate	80-62-6	113		0.7	IRIS	E									
Methyl tert-butyl ether	1634-04-4	114	3	3	IRIS		2.6E-07		CAL						
4,4'-Methylene bis(2-chloroaniline)	101-14-4	115	*1*			B2	0.00043		CAL						
Methylene chloride	75-09-2	116	2B	*0.6*	*IRIS*	*LH*	*1E-08*	*M**	IRIS						
Methylene diphenyl diisocyanate	101-68-8	117		0.0006	IRIS	InI									
4,4'-Methylenedianiline	101-77-9	118	2B	0.02	CAL		0.00046		CAL						
Naphthalene	91-20-3	119		0.003	IRIS	C	0.000034		CAL						
Nickel compounds	7440-02-0	186	*1*	0.00009	ATSDR	A									
Nickel oxide	1313-99-1	186													
Nickel refinery dust	NI_DUST	186				A	0.00024		IRIS						

Revisions since 04/27/10 are shown in red. **CAS NO.** = Chemical Abstracts Services number for the compound. **HAP NO.** = Position of the compound on the HAP list in the Clean Air Act (112[b][2]). "999" denotes substances under consideration for listing. **Sources:** IRIS = Integrated Risk Information System; ATSDR = US Agency for Toxic Substances and Disease Registry; D-ATSDR = draft ATSDR; CA = California EPA; P-CAL = Proposed CAL; HEAST = EPA Health Effects Assessment Tables				**IARC WOE** = weight of evidence for carcinogenicity in humans (1 - carcinogenic; 2A - probably carcinogenic; 2B - possibly carcinogenic; 3 - not classifiable; 4 - probably not carcinogenic).		**EPA WOE** (2005 Guidelines) = weight of evidence for carcinogenicity under 2005 EPA cancer guidelines: CH - carcinogenic to humans; LH - likely to be carcinogenic; SE - suggestive evidence of carcinogenic potential; InI - inadequate information to assess carcinogenic potential; NH - not likely to be carcinogenic). **EPA MOA** (2005 Guidelines) = mode of action for carcinogenicity: M* - mutagenic and early life data lacking; age-dependent adjustment factors should be applied when assessing risk for ages younger than 16 years per 2005 Supplemental Guidance. *[See Table4A-1 notes.]*						**EPA WOE** (1986 Guidelines) = weight of evidence for carcinogenicity under the 1986 EPA cancer guidelines: A - human carcinogen; B1 - probable carcinogen, limited human evidence; B2 - probable carcinogen, sufficient evidence in animals; C - possible human carcinogen; D - not classifiable; E - evidence of noncarcinogenicity.			
				CHRONIC INHALATION						CHRONIC ORAL					
				NONCANCER		CANCER				NONCANCER		CANCER			
CHEMICAL NAME	CAS NO.	HAP NO.	IARC WOE	mg/m³	SOURCE	EPA WOE	1/(ug/m³)	MOA	SOURCE	mg/kg/d	SOURCE	EPA WOE	1/(mg/kg/d)	MOA	SOURCE
Nickel subsulfide	12035-72-2	186				A	0.00048		IRIS						
Nitrobenzene	98-95-3	120	2B	0.009	IRIS	LH	0.00004		IRIS						
2-Nitropropane	79-46-9	123	2B	0.02	IRIS	B2	5.6E-06		EPA OAQPS						
Nitrosodimethylamine	62-75-9	125	2A			B2	0.014	M*	IRIS						
N-Nitrosomorpholine	59-89-2	126	2B				0.0019		CAL						
Parathion	56-38-2	127	3			C									
Polychlorinated biphenyls	1336-36-3	136	2A			B2	0.0001		IRIS			B2	2		IRIS
Aroclor 1016	12674-11-2	136								0.00007	IRIS				
Aroclor 1254	11097-69-1	136								0.00002	IRIS				
Pentachloronitrobenzene	82-68-8	128	3			C									
Pentachlorophenol	87-86-5	129	2B	0.1	P-CAL	B2	5.1E-06		CAL						
Phenol	108-95-2	130	3	0.2	CAL	InI									
p-Phenylenediamine	106-50-3	131													
Phosgene	75-44-5	132		0.0003	IRIS	InI									
Phosphine	7803-51-2	133		0.0003	IRIS	InI									
Phosphorus, white	7723-14-0	134		0.00007	P-CAL	D									
Phthalic anhydride	85-44-9	135		0.02	CAL										
Polybrominated biphenyls	59536-65-1	187	2B			B2				0.000007	HEAST	B2	8.9		HEAST
Acenaphthene	83-32-9	187	3			D				0.06	IRIS	D			
Acenaphthylene	206-96-8	187				D						D			
2-Aminoanthraquinone	117-79-3	187	3				9.4E-06		CAL				0.033		CAL

Revisions since 04/27/10 are shown in red.
CAS NO. = Chemical Abstracts Services number for the compound.
HAP NO. = Position of the compound on the HAP list in the Clean Air Act (112[b][2]). "999" denotes substances under consideration for listing.
Sources: IRIS = Integrated Risk Information System; ATSDR = US Agency for Toxic Substances and Disease Registry; D-ATSDR = draft ATSDR; CA = California EPA; P-CAL = Proposed CAL; HEAST = EPA Health Effects Assessment Tables

IARC WOE = weight of evidence for carcinogenicity in humans (1 - carcinogenic; 2A - probably carcinogenic; 2B - possibly carcinogenic; 3 - not classifiable; 4 - probably not carcinogenic).

EPA WOE (2005 Guidelines) = weight of evidence for carcinogenicity under 2005 EPA cancer guidelines: CH - carcinogenic to humans; LH - likely to be carcinogenic; SE - suggestive evidence of carcinogenic potential; InI - inadequate information to assess carcinogenic potential; NH - not likely to be carcinogenic).
EPA MOA (2005 Guidelines) = mode of action for carcinogenicity: M* - mutagenic and early life data lacking; age-dependent adjustment factors should be applied when assessing risk for ages younger than 16 years per 2005 Supplemental Guidance. *[See Table4A-1 notes.]*

EPA WOE (1986 Guidelines) = weight of evidence for carcinogenicity under the 1986 EPA cancer guidelines: A - human carcinogen; B1 - probable carcinogen, limited human evidence; B2 - probable carcinogen, sufficient evidence in animals; C - possible human carcinogen; D - not classifiable E - evidence of noncarcinogenicity.

				CHRONIC INHALATION						CHRONIC ORAL					
				NONCANCER		CANCER				NONCANCER		CANCER			
CHEMICAL NAME	CAS NO.	HAP NO.	IARC WOE	mg/m^3	SOURCE	EPA WOE	1/(ug/m^3)	MOA	SOURCE	mg/kg/d	SOURCE	EPA WOE	1/(mg/kg/d)	MOA	SOURCE
Anthracene	120-12-7	187	3			D				0.3	IRIS	D			
Benzo(a)anthracene	56-55-3	187	2B			B2	0.00011		CAL			B2	1.2		CAL
Benzo(b)fluoranthene	205-99-2	187	2B			B2	0.00011		CAL			B2	1.2		CAL
Benzo[j]fluoranthene	205-82-3	187	2B				0.00011		CAL				1.2		CAL
Benzo(k)fluoranthene	207-08-9	187	2B			B2	0.00011		CAL			B2	1.2		CAL
Benzo(g,h,i)perylene	191-24-2	187	3			D						D			
Benzo(a)pyrene	50-32-8	187	1			B2	0.0011	M*	CAL			B2	7.3	M*	IRIS
Benzo(e)pyrene	192-97-2	187	3												
Carbazole	86-74-8	187	2B			B2						B2	0.02		HEAST
beta-Chloronaphthalene	91-58-7	187								0.08	IRIS				
Chrysene	218-01-9	187	2B			B2	0.000011		CAL			B2	0.12		CAL
Dibenz[a,h]acridine	226-36-8	187	2B				0.00011		CAL				1.2		CAL
Dibenz[a,j]acridine	224-42-0	187	2A				0.00011		CAL				1.2		CAL
Dibenz(a,h)anthracene	53-70-3	187	2A			B2	0.0012	M*	CAL			B2	4.1	M*	CAL
7H-Dibenzo[c,g]carbazole	194-59-2	187	2B				0.0011		CAL				12		CAL
Dibenzo[a,e]pyrene	192-65-4	187	3				0.0011		CAL				12		CAL
Dibenzo[a,h]pyrene	189-64-0	187	2B				0.011		CAL				120		CAL
Dibenzo[a,i]pyrene	189-55-9	187	2B				0.011		CAL				120		CAL
Dibenzo[a,l]pyrene	191-30-0	187	2A				0.011		CAL				120		CAL
7,12-Dimethylbenz(a) anthracene	57-97-6	187					0.071	M*	CAL				250	M*	CAL
1,6-Dinitropyrene	42397-64-8	187	2B				0.011		CAL				120		CAL

Revisions since 04/27/10 are shown in red.
CAS NO. = Chemical Abstracts Services number for the compound.
HAP NO. = Position of the compound on the HAP list in the Clean Air Act (112[b][2]). "999" denotes substances under consideration for listing.
Sources: IRIS = Integrated Risk Information System; ATSDR = US Agency for Toxic Substances and Disease Registry; D-ATSDR = draft ATSDR; CA = California EPA; P-CAL = Proposed CAL; HEAST = EPA Health Effects Assessment Tables

IARC WOE = weight of evidence for carcinogenicity in humans (1 - carcinogenic; 2A - probably carcinogenic; 2B - possibly carcinogenic; 3 - not classifiable; 4 - probably not carcinogenic).

EPA WOE (2005 Guidelines) = weight of evidence for carcinogenicity under 2005 EPA cancer guidelines:
CH - carcinogenic to humans; LH - likely to be carcinogenic;
SE - suggestive evidence of carcinogenic potential;
InI - inadequate information to assess carcinogenic potential;
NH - not likely to be carcinogenic).
EPA MOA (2005 Guidelines) = mode of action for carcinogenicity:
M* - mutagenic and early life data lacking; age-dependent adjustment factors should be applied when assessing risk for ages younger than 16 years per 2005 Supplemental Guidance.
[See Table4A-1 notes.]

EPA WOE (1986 Guidelines) = weight of evidence for carcinogenicity under the 1986 EPA cancer guidelines:
A - human carcinogen;
B1 - probable carcinogen, limited human evidence;
B2 - probable carcinogen, sufficient evidence in animals;
C - possible human carcinogen; D - not classifiable
E - evidence of noncarcinogenicity.

				CHRONIC INHALATION						CHRONIC ORAL					
				NONCANCER		CANCER				NONCANCER		CANCER			
CHEMICAL NAME	CAS NO.	HAP NO.	IARC WOE	mg/m^3	SOURCE	EPA WOE	1/(ug/m^3)	MOA	SOURCE	mg/kg/d	SOURCE	EPA WOE	1/(mg/kg/d)	MOA	SOURCE
1,8-Dinitropyrene	42397-65-9	187	2B				0.0011		CAL				12		CAL
Fluoranthene	206-44-0	187	3			D				0.04	IRIS	D			
Fluorene	86-73-7	187	3			D				0.04	IRIS	D			
Indeno(1,2,3-cd)pyrene	193-39-5	187	2B			B2	0.00011		CAL			B2	1.2		CAL
3-Methylcholanthrene	56-49-5	187					0.0063	M*	CAL				22	M*	CAL
5-Methylchrysene	3697-24-3	187	2B				0.0011		CAL				12		CAL
1-Methylnaphthalene	90-12-0	187								0.07	ATSDR				
2-Methylnaphthalene	91-57-6	187				InI				0.04	ATSDR	InI			
5-Nitroacenaphthene	602-87-9	187	2B				0.000037		CAL				0.13		CAL
6-Nitrochrysene	7496-02-8	187	2B				0.011		CAL				120		CAL
2-Nitrofluorene	607-57-8	187	2B				0.000011		CAL				0.12		CAL
1-Nitropyrene	5522-43-0	187	2B				0.00011		CAL				1.2		CAL
4-Nitropyrene	57835-92-4	187	2B				0.00011		CAL				1.2		CAL
Octabromodiphenyl ether	32536-52-0	187				D				0.003	IRIS	D			
Phenanthrene	85-01-8	187	3			D						D			
Pyrene	129-00-0	187	3			D				0.03	IRIS	D			
1,3-Propane sultone	1120-71-4	137	2B				0.00069		CAL						
Propionaldehyde	123-38-6	139		0.008	IRIS	InI									
Propoxur	114-26-1	140				B2									
Propylene dichloride	78-87-5	141		0.004	IRIS	B2									
Propylene oxide	75-56-9	142	2B	0.03	IRIS	B2	3.7E-06		IRIS						

Revisions since 04/27/10 are shown in red.
CAS NO. = Chemical Abstracts Services number for the compound.
HAP NO. = Position of the compound on the HAP list in the Clean Air Act (112[b][2]). "999" denotes substances under consideration for listing.
Sources: IRIS = Integrated Risk Information System; ATSDR = US Agency for Toxic Substances and Disease Registry; D-ATSDR = draft ATSDR; CA = California EPA; P-CAL = Proposed CAL; HEAST = EPA Health Effects Assessment Tables

IARC WOE = weight of evidence for carcinogenicity in humans (1 - carcinogenic; 2A - probably carcinogenic; 2B - possibly carcinogenic; 3 - not classifiable; 4 - probably not carcinogenic).

EPA WOE (2005 Guidelines) = weight of evidence for carcinogenicity under 2005 EPA cancer guidelines:
CH - carcinogenic to humans; LH - likely to be carcinogenic;
SE - suggestive evidence of carcinogenic potential;
InI - inadequate information to assess carcinogenic potential;
NH - not likely to be carcinogenic).
EPA MOA (2005 Guidelines) = mode of action for carcinogenicity:
M* - mutagenic and early life data lacking; age-dependent adjustment factors should be applied when assessing risk for ages younger than 16 years per 2005 Supplemental Guidance. *[See Table4A-1 notes.]*

EPA WOE (1986 Guidelines) = weight of evidence for carcinogenicity under the 1986 EPA cancer guidelines:
A - human carcinogen;
B1 - probable carcinogen, limited human evidence;
B2 - probable carcinogen, sufficient evidence in animals;
C - possible human carcinogen; D - not classifiable
E - evidence of noncarcinogenicity.

				CHRONIC INHALATION						CHRONIC ORAL					
				NONCANCER		CANCER				NONCANCER		CANCER			
CHEMICAL NAME	CAS NO.	HAP NO.	IARC WOE	mg/m^3	SOURCE	EPA WOE	1/(ug/m^3)	MOA	SOURCE	mg/kg/d	SOURCE	EPA WOE	1/(mg/kg/d)	MOA	SOURCE
Quinoline	91-22-5	144				LH									
Selenium compounds	7782-49-2	189	3	0.02	CAL	D									
Hydrogen selenide	7783-07-5	189		0.00008	P-CAL										
Selenious acid	7783-00-8	189				D									
Selenium dioxide	7446-08-4	189		0.02	CAL										
Selenium disulfide	7488-56-4	189		0.02	CAL										
Selenium sulfide	7446-34-6	189		0.02	CAL	B2									
Selenourea	630-10-4	189													
Styrene	100-42-5	146	2B	1	IRIS										
Styrene oxide	96-09-3	147	2A	0.006	P-CAL										
2,3,7,8-Tetrachlorodibenzo-p-dioxin	1746-01-6	148	1	4E-08	CAL	B2	33		EPA ORD	*7E-10*	*IRIS*	B2	150000		EPA ORD
1,1,2,2-Tetrachloroethane	79-34-5	149	3			C	0.000058		IRIS						
Tetrachloroethene	127-18-4	150	2A	*0.04*	*IRIS*	*LH*	*2.6E-07*		*IRIS*						
Titanium tetrachloride	7550-45-0	151		0.0001	ATSDR										
Toluene	108-88-3	152	3	5	IRIS	InI									
2,4-Toluene diamine	95-80-7	153	2B			B2	0.0011		CAL						
2,4/2,6-Toluene diisocyanate mixture (TDI)	26471-62-5	154	2B	0.00007	IRIS		0.000011		CAL						
o-Toluidine	95-53-4	155	*1*			B2	0.000051		CAL						

Revisions since 04/27/10 are shown in red.
CAS NO. = Chemical Abstracts Services number for the compound.
HAP NO. = Position of the compound on the HAP list in the Clean Air Act (112[b][2]). "999" denotes substances under consideration for listing.
Sources: IRIS = Integrated Risk Information System; ATSDR = US Agency for Toxic Substances and Disease Registry; D-ATSDR = draft ATSDR; CA = California EPA; P-CAL = Proposed CAL; HEAST = EPA Health Effects Assessment Tables

IARC WOE = weight of evidence for carcinogenicity in humans (1 - carcinogenic; 2A - probably carcinogenic; 2B - possibly carcinogenic; 3 - not classifiable; 4 - probably not carcinogenic).

EPA WOE (2005 Guidelines) = weight of evidence for carcinogenicity under 2005 EPA cancer guidelines:
CH - carcinogenic to humans; LH - likely to be carcinogenic;
SE - suggestive evidence of carcinogenic potential;
InI - inadequate information to assess carcinogenic potential;
NH - not likely to be carcinogenic).
EPA MOA (2005 Guidelines) = mode of action for carcinogenicity:
M* - mutagenic and early life data lacking; age-dependent adjustment factors should be applied when assessing risk for ages younger than 16 years per 2005 Supplemental Guidance. *[See Table4A-1 notes.]*

EPA WOE (1986 Guidelines) = weight of evidence for carcinogenicity under the 1986 EPA cancer guidelines:
A - human carcinogen;
B1 - probable carcinogen, limited human evidence;
B2 - probable carcinogen, sufficient evidence in animals;
C - possible human carcinogen; D - not classifiable
E - evidence of noncarcinogenicity.

				CHRONIC INHALATION						CHRONIC ORAL					
				NONCANCER		CANCER				NONCANCER		CANCER			
CHEMICAL NAME	CAS NO.	HAP NO.	IARC WOE	mg/m^3	SOURCE	EPA WOE	1/(ug/m^3)	MOA	SOURCE	mg/kg/d	SOURCE	EPA WOE	1/(mg/kg/d)	MOA	SOURCE
Toxaphene	8001-35-2	156	2B			B2	0.00032		IRIS			B2	1.1		IRIS
1,2,4-Trichlorobenzene	120-82-1	157		0.2	HEAST	D									
1,1,2-Trichloroethane	79-00-5	158	3	0.4	P-CAL	C	0.000016		IRIS						
Trichloroethylene	79-01-6	159	2A	0.002	IRIS	CH	4.1E-06	M*	IRIS						
2,4,5-Trichlorophenol	95-95-4	160													
2,4,6-Trichlorophenol	88-06-2	161				B2	3.1E-06		IRIS						
Triethylamine	121-44-8	162		0.007	IRIS										
Trifluralin	1582-09-8	163	3			C				0.0075	IRIS	C	0.0077		IRIS
Uranium compounds	7440-61-1	188													
Uranium, insoluble salts	URANINSOLS	188		0.0008	D-ATSDR										
Uranium, soluble salts	URANSOLS	188		0.00004	D-ATSDR										
Vinyl acetate	108-05-4	165	2B	0.2	IRIS										
Vinyl bromide	593-60-2	166	2A	0.003	IRIS	B2	0.000032		HEAST						
Vinyl chloride	75-01-4	167	1	0.1	IRIS	CH	8.8E-06		IRIS						
Vinylidene chloride	75-35-4	168	3	0.2	IRIS	SE									
m-Xylene	108-38-3	171													
o-Xylene	95-47-6	170													
Xylenes (mixed)	1330-20-7	169	3	0.1	IRIS	InI									

Source: USEPA

Table A-2 Acute dose-response values for screening risk assessments (12/19/2011)

AEGL = Acute exposure guideline levels for mild effects (AEGL-1) and moderate effects (AEGL-2) for 1- and 8-hour exposures. Superscripts indicate the AEGL's status: f = final, i=interim, and p=proposed. **ERPG** = US DOE Emergency Removal Program guidelines for mild or transient effects (ERPG-1) and irreversible or serious effects (ERPG-2) for 1-hour exposures. **MRL** = ATSDR minimum risk levels for no adverse effects for 1 to 14-day exposures. **REL** = California EPA reference exposure level for no adverse effects. Most, but not all, RELs are for 1-hour exposures. **IDLH/10** = One-tenth of levels determined by NIOSH to be imminently dangerous to life and health, approximately comparable to mild effects levels for 1-hour exposures. **TEEL** = US DOE Temporary emergency exposure limits for no effects (TEEL-0) and mild, transient effects (TEEL-1) for 1-hour exposures. *TEELs are derived according to a tiered, formula-like methodology, and do not undergo peer review. They are not recommended as the basis for regulatory decision-making and are shown here only to inform situations where acute values from other sources are not available.*

			AEGL-1 (1-h)	AEGL-1 (8-h)	AEGL-2 (1-h)	AEGL-2 (8-h)	ERPG-1	ERPG-2	MRL	REL	IDLH/10	TEEL-0	TEEL-1
CHEMICAL NAME	CAS NO.	HAP NO.	mg/m^3	mg/m^3	mg/m^3	mg/m^3	mg/m^3	mg/m^3	mg/m^3	mg/m^3	mg/m^3	mg/m^3	mg/m^3
Acetaldehyde	75-07-0	1	81^{i}	81^{i}	490^{i}	200^{i}	18	360		0.47	360		
Acetamide	60-35-5	2										25	75
Acetonitrile	75-05-8	3	22^{i}	22^{i}	540^{i}	140^{i}					84		
Acetophenone	98-86-2	4										10	30
2-Acetylaminofluorene	53-96-3	5										0.25	0.75
Acrolein	107-02-8	6	0.069^{i}	0.069^{i}	0.23^{i}	0.23^{i}	0.11	0.34	0.0069	0.0025	0.46		
Acrylamide	79-06-1	7									6		
Acrylic acid	79-10-7	8	4.4^{i}	4.4^{i}	140^{i}	41^{i}	2.9	150		6			
Acrylonitrile	107-13-1	9	10^{i}	10^{i}	130^{i}	19^{i}	22	77	0.22		19		
Allyl chloride	107-05-1	10	8.8^{i}	8.8^{i}	170^{i}	69^{i}	9.4	130			78		
4-Aminobiphenyl	92-67-1	11										0.5	1.5
Aniline	62-53-3	12	30^{f}	3.8^{f}	46^{f}	5.7^{f}					38		
Anisidine	90-04-0	13									5		
Antimony compounds	7440-36-0	173									5		
Antimony pentafluoride	7783-70-2	173										0.75	0.75
Antimony potassium tartrate	304-61-0	173										1.2	4
Antimony trihydride	7803-52-3	173		^{i}	7.7^{i}	0.92^{i}							
Antimony trioxide	1309-64-4	173										0.6	1.5
Arsenic chloride	7784-34-1	174										0.19	0.56
Arsenic compounds	7440-38-2	174								0.0002	0.5		
Arsenic oxide	1327-53-3	174		^{i}	3^{i}	1.2^{i}							
Arsenic pentoxide	1303-28-2	174										0.015	0.045
Arsine	7784-42-1	174		^{f}	0.54^{f}	0.064^{f}		1.6			0.96		
Benzene	71-43-2	15	170^{i}	29^{i}	2600^{i}	640^{i}	160	480	0.029	1.3	160		

AEGL = Acute exposure guideline levels for mild effects (AEGL-1) and moderate effects (AEGL-2) for 1- and 8-hour exposures. Superscripts indicate the AEGL's status: f = final, i=interim, and p=proposed. **ERPG** = US DOE Emergency Removal Program guidelines for mild or transient effects (ERPG-1) and irreversible or serious effects (ERPG-2) for 1-hour exposures. **MRL** = ATSDR minimum risk levels for no adverse effects for 1 to 14-day exposures. **REL** = California EPA reference exposure level for no adverse effects. Most, but not all, RELs are for 1-hour exposures. **IDLH/10** = One-tenth of levels determined by NIOSH to be imminently dangerous to life and health, approximately comparable to mild effects levels for 1-hour exposures. **TEEL** = US DOE Temporary emergency exposure limits for no effects (TEEL-0) and mild, transient effects (TEEL-1) for 1-hour exposures. *TEELs are derived according to a tiered, formula-like methodology, and do not undergo peer review. They are not recommended as the basis for regulatory decision-making and are shown here only to inform situations where acute values from other sources are not available.*

			AEGL-1 (1-h)	AEGL-1 (8-h)	AEGL-2 (1-h)	AEGL-2 (8-h)	ERPG-1	ERPG-2	MRL	REL	IDLH/10	TEEL-0	TEEL-1
CHEMICAL NAME	CAS NO.	HAP NO.	mg/m³	mg/m³	mg/m³	mg/m³	mg/m³	mg/m³	mg/m³	mg/m³	mg/m³	mg/m³	mg/m³
Benzidine	92-87-5	16										0.15	0.5
Benzotrichloride	98-07-7	17										0.1	0.1
Benzyl chloride	100-44-7	18					5.2	52		0.24	5.2		
Beryllium chloride	7787-47-5	175										0.015	0.04
Beryllium compounds	7440-41-7	175						0.025			0.4		
Beryllium fluoride	7787-49-7	175										0.01	0.025
Beryllium nitrate	13597-99-4	175										0.03	0.075
Beryllium oxide	1304-56-9	175										0.005	
Biphenyl	92-52-4	19		[i]	61[i]	28[i]							
Bis(2-ethylhexyl)phthalate	117-81-7	20										5	10
Bis(chloromethyl)ether	542-88-1	21		[i]	0.21[i]	0.094[i]		0.47					
Bromoform	75-25-2	22									880		
1,3-Butadiene	106-99-0	23	1500[i]	1500[i]	12000[i]	6000[i]	22	440	0.22		440		
Cadmium compounds	7440-43-9	176							0.00003		0.9		
Cadmium oxide	1306-19-0	176										0.005	0.035
Cadmium stearate	2223-93-0	176										0.03	0.15
Calcium cyanamide	156-62-7	24										0.5	1.5
Captan	133-06-2	26										5	15
Carbaryl	63-25-2	27									10		
Carbon disulfide	75-15-0	28	40[f]	21[f]	500[f]	160[f]	3.1	500		6.2	160		
Carbon tetrachloride	56-23-5	29	280[i]	120[i]	1200[i]	510[i]	130	630		1.9	130		
Carbonyl sulfide	463-58-1	30		[i]	140[i]	57[i]							
Catechol	120-80-9	31										23	68
Chloramben	133-90-4	32										35	100

AEGL = Acute exposure guideline levels for mild effects (AEGL-1) and moderate effects (AEGL-2) for 1- and 8-hour exposures. Superscripts indicate the AEGL's status: f = final, i=interim, and p=proposed. **ERPG** = US DOE Emergency Removal Program guidelines for mild or transient effects (ERPG-1) and irreversible or serious effects (ERPG-2) for 1-hour exposures. **MRL** = ATSDR minimum risk levels for no adverse effects for 1 to 14-day exposures. **REL** = California EPA reference exposure level for no adverse effects. Most, but not all, RELs are for 1-hour exposures. **IDLH/10** = One-tenth of levels determined by NIOSH to be imminently dangerous to life and health, approximately comparable to mild effects levels for 1-hour exposures. **TEEL** = US DOE Temporary emergency exposure limits for no effects (TEEL-0) and mild, transient effects (TEEL-1) for 1-hour exposures. *TEELs are derived according to a tiered, formula-like methodology, and do not undergo peer review. They are not recommended as the basis for regulatory decision-making and are shown here only to inform situations where acute values from other sources are not available.*

			AEGL-1 (1-h)	AEGL-1 (8-h)	AEGL-2 (1-h)	AEGL-2 (8-h)	ERPG-1	ERPG-2	MRL	REL	IDLH/10	TEEL-0	TEEL-1
CHEMICAL NAME	CAS NO.	HAP NO.	mg/m^3	mg/m^3	mg/m^3	mg/m^3	mg/m^3	mg/m^3	mg/m^3	mg/m^3	mg/m^3	mg/m^3	mg/m^3
Chlordane	57-74-9	33									10		
Chlorine	7782-50-5	34	1.5^{f}	1.5^{f}	5.8^{f}	2.1^{f}	2.9	8.7	0.2	0.21	2.9		
Chloroacetic acid	79-11-8	35		^{f}	26^{f}	3.2^{f}							
2-Chloroacetophenone	532-27-4	36											
Chlorobenzene	108-90-7	37	46^{i}	46^{i}	690^{i}	690^{i}					460		
Chlorobenzilate	510-15-6	38										0.075	0.25
Chloroform	67-66-3	39		^{i}	310^{i}	140^{i}		240	0.49	0.15	240		
Chloromethyl methyl ether	107-30-2	40		^{i}	1.6^{i}	0.73^{i}		3.3					
Chloroprene	126-99-8	41									110		
Chromium (III) compounds	16065-83-1	177											
Chromium (VI) compounds	18540-29-9	177									1.5		
Chromium (VI) trioxide, chromic acid mist	11115-74-5	177									1.5		
Chromium chloride	10025-73-7	177										1.5	4
Chromium compounds	7440-47-3	177										1	1.5
Cobalt bromide	7789-43-7	178										0.2	0.2
Cobalt carbonate	513-79-1	178										0.12	0.12
Cobalt carbonyl	10210-68-1	178										0.27	0.27
Cobalt chloride	7646-79-9	178										0.12	0.12
Cobalt compounds	7440-48-4	178									2		
Cobalt hydrocarbonyl	16842-03-8	178						0.9					
Cobalt nitrate	Co Nitrate	178										0.15	0.15
Cobalt oxides (mixed)	COBOXIDES	178										0.075	0.075
Coke Oven Emissions	8007-45-2	179										0.1	1.2
m-Cresol	108-39-4	44									110		

AEGL = Acute exposure guideline levels for mild effects (AEGL-1) and moderate effects (AEGL-2) for 1- and 8-hour exposures. Superscripts indicate the AEGL's status: f = final, i=interim, and p=proposed. **ERPG** = US DOE Emergency Removal Program guidelines for mild or transient effects (ERPG-1) and irreversible or serious effects (ERPG-2) for 1-hour exposures. **MRL** = ATSDR minimum risk levels for no adverse effects for 1 to 14-day exposures. **REL** = California EPA reference exposure level for no adverse effects. Most, but not all, RELs are for 1-hour exposures. **IDLH/10** = One-tenth of levels determined by NIOSH to be imminently dangerous to life and health, approximately comparable to mild effects levels for 1-hour exposures. **TEEL** = US DOE Temporary emergency exposure limits for no effects (TEEL-0) and mild, transient effects (TEEL-1) for 1-hour exposures. *TEELs are derived according to a tiered, formula-like methodology, and do not undergo peer review. They are not recommended as the basis for regulatory decision-making and are shown here only to inform situations where acute values from other sources are not available.*

			AEGL-1 (1-h)	AEGL-1 (8-h)	AEGL-2 (1-h)	AEGL-2 (8-h)	ERPG-1	ERPG-2	MRL	REL	IDLH/10	TEEL-0	TEEL-1
CHEMICAL NAME	CAS NO.	HAP NO.	mg/m^3	mg/m^3	mg/m^3	mg/m^3	mg/m^3	mg/m^3	mg/m^3	mg/m^3	mg/m^3	mg/m^3	mg/m^3
o-Cresol	95-48-7	43									110		
p-Cresol	106-44-5	45									110		
Cresols (mixed)	1319-77-3	42									110		
Cumene	98-82-8	46	250^{i}	250^{i}	1500^{i}	640^{i}					440		
Cyanophos	2636-26-2	180										1.2	3.5
Cyanuric fluoride	675-14-9	180										0.17	0.17
Cyanide compounds	57-12-5	180									2.5		
Acetone cyanohydrin	75-86-5	180	7^{f}	3.5^{f}	25^{f}	8.7^{f}							
Barium cyanide	542-62-1	180										0.6	2
Calcium cyanide	592-01-8	180	3.8^{i}	1.9^{i}	13^{i}	4.7^{i}							
Copper cyanide	544-92-3	180										1.2	4
Cyanogen	460-19-5	180	4.3^{i}	2.1^{i}	18^{i}	9.2^{i}							
Cyanogen bromide	506-68-3	180										20	44
Cyanogen chloride	506-77-4	180						1					
Cyanogen iodide	506-78-5	180										35	100
Hydrogen cyanide	74-90-8	180	2.2^{f}	1.1^{f}	7.8^{f}	2.8^{f}		11		0.34	5.5		
Potassium cyanide	151-50-8	180										5	5
Potassium silver cyanide	506-61-6	180										1	3
Potassium thiocyanate	333-20-0	180										10	35
Silver cyanide	506-64-9	180										25	25
Sodium cyanide	143-33-9	180										5	5
Zinc cyanide	557-21-1	180										20	20
2,4-D, salts and esters	94-75-7	47									10		
DDE	72-55-9	48										10	30

AEGL = Acute exposure guideline levels for mild effects (AEGL-1) and moderate effects (AEGL-2) for 1- and 8-hour exposures. Superscripts indicate the AEGL's status: f = final, i=interim, and p=proposed. **ERPG** = US DOE Emergency Removal Program guidelines for mild or transient effects (ERPG-1) and irreversible or serious effects (ERPG-2) for 1-hour exposures. **MRL** = ATSDR minimum risk levels for no adverse effects for 1 to 14-day exposures. **REL** = California EPA reference exposure level for no adverse effects. Most, but not all, RELs are for 1-hour exposures. **IDLH/10** = One-tenth of levels determined by NIOSH to be imminently dangerous to life and health, approximately comparable to mild effects levels for 1-hour exposures. **TEEL** = US DOE Temporary emergency exposure limits for no effects (TEEL-0) and mild, transient effects (TEEL-1) for 1-hour exposures. *TEELs are derived according to a tiered, formula-like methodology, and do not undergo peer review. They are not recommended as the basis for regulatory decision-making and are shown here only to inform situations where acute values from other sources are not available.*

			AEGL-1 (1-h)	AEGL-1 (8-h)	AEGL-2 (1-h)	AEGL-2 (8-h)	ERPG-1	ERPG-2	MRL	REL	IDLH/10	TEEL-0	TEEL-1
CHEMICAL NAME	CAS NO.	HAP NO.	mg/m^3	mg/m^3	mg/m^3	mg/m^3	mg/m^3	mg/m^3	mg/m^3	mg/m^3	mg/m^3	mg/m^3	mg/m^3
Diazomethane	334-88-3	49										0.34	1
Dibenzofuran	132-64-9	50										10	30
2,3,4,7,8-Pentachlorodibenzofuran	57117-31-4	187										0.000025	0.000075
1,2-Dibromo-3-chloropropane	96-12-8	51										0.0097	0.029
Dibutylphthalate	84-74-2	52									400		
p-Dichlorobenzene	106-46-7	53							12		90		
3,3'-Dichlorobenzidine	91-94-1	54										2.1	6.2
Dichloroethyl ether	111-44-4	55									58		
1,3-Dichloropropene	542-75-6	56										4.5	14
Dichlorvos	62-73-7	57	0.99^{p}	0.99^{p}	5.1^{p}	5.1^{p}			0.018		10		
Diesel engine emissions	DIESEL EMIS.	999										35	100
Diethanolamine	111-42-2	58										2	6
Diethyl sulfate	64-67-5	60										1.9	4.7
N,N-diethyl/dimethylaniline	Dialks	59											
3,3'-Dimethoxybenzidine	119-90-4	61										1.5	5
p-Dimethylaminoazobenzene	60-11-7	62										15	50
3,3'-Dimethylbenzidine	119-93-7	63										0.1	0.3
Dimethyl carbamoyl chloride	79-44-7	64										0.88	2.6
Dimethyl formamide	68-12-2	65		^{i}	270^{i}	110^{i}	6	300			150		
Dimethyl phthalate	131-11-3	67									200		
Dimethyl sulfate	77-78-1	68	0.12^{i}	0.045^{i}	0.62^{i}	0.22^{i}					3.6		
N,N-dimethylaniline	121-69-7	59									50		
1,1-Dimethylhydrazine	57-14-7	66		^{f}	7.4^{f}	0.93^{f}					3.7		
4,6-Dinitro-o-cresol	534-52-1	69									0.5		

AEGL = Acute exposure guideline levels for mild effects (AEGL-1) and moderate effects (AEGL-2) for 1- and 8-hour exposures. Superscripts indicate the AEGL's status: f = final, i=interim, and p=proposed. **ERPG** = US DOE Emergency Removal Program guidelines for mild or transient effects (ERPG-1) and irreversible or serious effects (ERPG-2) for 1-hour exposures. **MRL** = ATSDR minimum risk levels for no adverse effects for 1 to 14-day exposures. **REL** = California EPA reference exposure level for no adverse effects. Most, but not all, RELs are for 1-hour exposures. **IDLH/10** = One-tenth of levels determined by NIOSH to be imminently dangerous to life and health, approximately comparable to mild effects levels for 1-hour exposures. **TEEL** = US DOE Temporary emergency exposure limits for no effects (TEEL-0) and mild, transient effects (TEEL-1) for 1-hour exposures. *TEELs are derived according to a tiered, formula-like methodology, and do not undergo peer review. They are not recommended as the basis for regulatory decision-making and are shown here only to inform situations where acute values from other sources are not available.*

			AEGL-1 (1-h)	AEGL-1 (8-h)	AEGL-2 (1-h)	AEGL-2 (8-h)	ERPG-1	ERPG-2	MRL	REL	IDLH/10	TEEL-0	TEEL-1
CHEMICAL NAME	CAS NO.	HAP NO.	mg/m^3	mg/m^3	mg/m^3	mg/m^3	mg/m^3	mg/m^3	mg/m^3	mg/m^3	mg/m^3	mg/m^3	mg/m^3
2,4-dinitrophenol	51-28-5	70										3	7.5
2,4-Dinitrotoluene	121-14-2	71									5		
2,4/2,6-Dinitrotoluene (mixture)	25321-14-6	71										0.2	0.6
1,4-Dioxane	123-91-1	72	61^{i}	61^{i}	1200^{i}	360^{i}			7.2	3	180		
1,2-Diphenylhydrazine	122-66-7	73										10	30
Epichlorohydrin	106-89-8	74	22^{i}	22^{i}	91^{i}	38^{i}	19	76		1.3	28		
1,2-Epoxybutane	106-88-7	75	210^{i}	210^{i}	410^{i}	410^{i}							
Ethyl acrylate	140-88-5	76	34^{i}	34^{i}	150^{i}	38^{i}	0.041	120			120		
Ethyl benzene	100-41-4	77	140^{i}	140^{i}	4800^{i}	2500^{i}			22		350		
Ethyl carbamate	51-79-6	78										500	500
Ethyl chloride	75-00-3	79							40		1000		
Ethylene dibromide	106-93-4	80	130^{i}	35^{i}	180^{i}	50^{i}					77		
Ethylene dichloride	107-06-2	81					200	810			20		
Ethylene glycol	107-21-1	82							2				
Ethylene imine (Aziridine)	151-56-4	83		^{f}	8.1^{f}	0.83^{f}							
Ethylene oxide	75-21-8	84		^{f}	81^{f}	14^{f}		90			140		
Ethylene thiourea	96-45-7	85										3.5	10
Ethylidene dichloride (1,1-Dichloroethane)	75-34-3	86									1200		
Formaldehyde	50-00-0	87	1.1^{f}	1.1^{f}	17^{f}	17^{f}	1.2	12	0.049	0.055	2.5		
Diethylene glycol monobutyl ether	112-34-5	181										100	150
Diethylene glycol monoethyl ether	111-90-0	181										140	410
Ethylene glycol ethyl ether	110-80-5	181								0.37	180		
Ethylene glycol ethyl ether acetate	111-15-9	181								0.14			
Ethylene glycol methyl ether	109-86-4	181								0.093			

AEGL = Acute exposure guideline levels for mild effects (AEGL-1) and moderate effects (AEGL-2) for 1- and 8-hour exposures. Superscripts indicate the AEGL's status: f = final, i=interim, and p=proposed. **ERPG** = US DOE Emergency Removal Program guidelines for mild or transient effects (ERPG-1) and irreversible or serious effects (ERPG-2) for 1-hour exposures. **MRL** = ATSDR minimum risk levels for no adverse effects for 1 to 14-day exposures. **REL** = California EPA reference exposure level for no adverse effects. Most, but not all, RELs are for 1-hour exposures. **IDLH/10** = One-tenth of levels determined by NIOSH to be imminently dangerous to life and health, approximately comparable to mild effects levels for 1-hour exposures. **TEEL** = US DOE Temporary emergency exposure limits for no effects (TEEL-0) and mild, transient effects (TEEL-1) for 1-hour exposures. *TEELs are derived according to a tiered, formula-like methodology, and do not undergo peer review. They are not recommended as the basis for regulatory decision-making and are shown here only to inform situations where acute values from other sources are not available.*

			AEGL-1 (1-h)	AEGL-1 (8-h)	AEGL-2 (1-h)	AEGL-2 (8-h)	ERPG-1	ERPG-2	MRL	REL	IDLH/10	TEEL-0	TEEL-1
CHEMICAL NAME	CAS NO.	HAP NO.	mg/m^3	mg/m^3	mg/m^3	mg/m^3	mg/m^3	mg/m^3	mg/m^3	mg/m^3	mg/m^3	mg/m^3	mg/m^3
Ethylene glycol methyl ether acetate	110-49-6	181											
Heptachlor	76-44-8	88									3.5		
Hexachlorobenzene	118-74-1	89										0.002	0.006
Hexachlorobutadiene	87-68-3	90					11	32					
Hexachlorocyclopentadiene	77-47-4	91										0.11	0.2
Hexachlorodibenzo-p-dioxin, mixture	19408-74-3	187										0.005	0.015
Hexachloroethane	67-72-1	92							58				
Hexamethylene-1,6-diisocyanate	822-06-0	93										0.034	0.1
Hexamethylphosphoramide	680-31-9	94										0.29	0.92
n-Hexane	110-54-3	95		^{i}	12000^{i}	12000^{i}					390		
Hydrazine	302-01-2	96	0.13^{i}	0.13^{f}	17^{f}	2.1^{f}	0.65	6.5			6.5		
Hydrochloric acid	7647-01-0	97	2.7^{f}	2.7^{f}	33^{f}	16^{f}	4.5	30		2.1	7.5		
Hydrofluoric acid	7664-39-3	98	0.82^{f}	0.82^{f}	20^{f}	9.8^{f}	1.6	16	0.016	0.24	2.5		
Hydrogen sulfide	7783-06-4	999	0.71^{i}	0.46^{i}	38^{i}	24^{i}	0.14	42	0.098	0.042			
Hydroquinone	123-31-9	99									5		
Isophorone	78-59-1	100										28	28
Lead acetate	301-04-2	182										0.075	0.2
Lead chloride	7758-95-4	182										0.06	0.2
Lead compounds	7439-92-1	182									10		
Lead nitrate	10099-74-8	182										0.075	0.22
Lead subacetate	1335-32-6	182										0.06	0.2
Tetraethyl lead	78-00-2	182									4		
Tetramethyl lead	75-74-1	182									4		
Lindane (gamma-HCH)	58-89-9	101									5		

AEGL = Acute exposure guideline levels for mild effects (AEGL-1) and moderate effects (AEGL-2) for 1- and 8-hour exposures. Superscripts indicate the AEGL's status: f = final, i=interim, and p=proposed. **ERPG** = US DOE Emergency Removal Program guidelines for mild or transient effects (ERPG-1) and irreversible or serious effects (ERPG-2) for 1-hour exposures. **MRL** = ATSDR minimum risk levels for no adverse effects for 1 to 14-day exposures. **REL** = California EPA reference exposure level for no adverse effects. Most, but not all, RELs are for 1-hour exposures. **IDLH/10** = One-tenth of levels determined by NIOSH to be imminently dangerous to life and health, approximately comparable to mild effects levels for 1-hour exposures. **TEEL** = US DOE Temporary emergency exposure limits for no effects (TEEL-0) and mild, transient effects (TEEL-1) for 1-hour exposures. *TEELs are derived according to a tiered, formula-like methodology, and do not undergo peer review. They are not recommended as the basis for regulatory decision-making and are shown here only to inform situations where acute values from other sources are not available.*

			AEGL-1 (1-h)	AEGL-1 (8-h)	AEGL-2 (1-h)	AEGL-2 (8-h)	ERPG-1	ERPG-2	MRL	REL	IDLH/10	TEEL-0	TEEL-1
CHEMICAL NAME	CAS NO.	HAP NO.	mg/m^3	mg/m^3	mg/m^3	mg/m^3	mg/m^3	mg/m^3	mg/m^3	mg/m^3	mg/m^3	mg/m^3	mg/m^3
alpha-Hexachlorocyclohexane (a-HCH)	319-84-6	101										0.5	1.5
beta-Hexachlorocyclohexane (b-HCH)	319-85-7	101										0.5	1.5
technical Hexachlorocyclohexane (HCH)	608-73-1	101										0.15	0.5
Maleic anhydride	108-31-6	102					0.8	8			1		
Manganese chloride	7773-01-5	183										0.4	6
Manganese compounds	7439-96-5	183									50		
Manganese dioxide	1313-13-9	183										0.3	4
Manganese oxide	1317-35-7	183										0.25	0.75
Manganese sulfate	7785-87-7	183										0.5	7.5
Manganese tricarbonyl methylcyclopentadienyl	12108-13-3	183										0.6	0.6
Mercuric acetate	1600-27-7	184										0.01	0.03
Mercuric chloride	7487-94-7	184										0.035	0.12
Mercuric nitrate	10045-94-0	184										0.04	0.15
Mercuric oxide	21908-53-2	184										0.025	0.1
Mercury (elemental)	7439-97-6	184		^{i}	1.7^{i}	0.33^{i}		2		0.0006			
Methylmercuric dicyanamide	502-39-6	184										0.015	0.04
Mercury compounds	HG_CMPDS	184									1		
Methoxyethylmercuric acetate	151-38-2	184										0.015	0.05
Methyl mercury	22967-92-6	184									0.2		
Phenylmercuric acetate	62-38-4	184										0.1	0.1
Methanol	67-56-1	103	690^{i}	350^{i}	2700^{i}	680^{i}	260	1300		28	790		

AEGL = Acute exposure guideline levels for mild effects (AEGL-1) and moderate effects (AEGL-2) for 1- and 8-hour exposures. Superscripts indicate the AEGL's status: f = final, i=interim, and p=proposed. **ERPG** = US DOE Emergency Removal Program guidelines for mild or transient effects (ERPG-1) and irreversible or serious effects (ERPG-2) for 1-hour exposures. **MRL** = ATSDR minimum risk levels for no adverse effects for 1 to 14-day exposures. **REL** = California EPA reference exposure level for no adverse effects. Most, but not all, RELs are for 1-hour exposures. **IDLH/10** = One-tenth of levels determined by NIOSH to be imminently dangerous to life and health, approximately comparable to mild effects levels for 1-hour exposures. **TEEL** = US DOE Temporary emergency exposure limits for no effects (TEEL-0) and mild, transient effects (TEEL-1) for 1-hour exposures. *TEELs are derived according to a tiered, formula-like methodology, and do not undergo peer review. They are not recommended as the basis for regulatory decision-making and are shown here only to inform situations where acute values from other sources are not available.*

			AEGL-1 (1-h)	AEGL-1 (8-h)	AEGL-2 (1-h)	AEGL-2 (8-h)	ERPG-1	ERPG-2	MRL	REL	IDLH/10	TEEL-0	TEEL-1
CHEMICAL NAME	CAS NO.	HAP NO.	mg/m^3	mg/m^3	mg/m^3	mg/m^3	mg/m^3	mg/m^3	mg/m^3	mg/m^3	mg/m^3	mg/m^3	mg/m^3
Methoxychlor	72-43-5	104									500		
Methyl bromide	74-83-9	105		^{i}	820^{i}	260^{i}		190	0.19	3.9	97		
Methyl chloride	74-87-3	106		^{i}	1900^{i}	780^{i}		830	1		410		
Methyl chloroform (1,1,1-Trichloroethane)	71-55-6	107	1300^{i}	1300^{i}	3300^{i}	1700^{i}	1900	3800	11	68	380		
Methyl hydrazine	60-34-4	109		^{f}	3.2^{f}	0.39^{f}					7.2		
Methyl iodide	74-88-4	110					150	290			58		
Methyl isobutyl ketone	108-10-1	111										310	310
Methyl isocyanate	624-83-9	112		^{f}	0.16^{f}	0.019^{f}	0.058	0.58			0.7		
Methyl methacrylate	80-62-6	113	70^{i}	70^{i}	490^{i}	200^{i}					410		
Methyl tert-butyl ether	1634-04-4	114	180^{i}	180^{i}	2100^{i}	1400^{i}			7.2				
4,4'-Methylene bis(2-chloroaniline)	101-14-4	115										0.11	0.33
Methylene chloride	75-09-2	116	690^{i}	^{i}	1900^{i}	210^{i}	1000	2600	2.1	14	800		
Methylene diphenyl diisocyanate	101-68-8	117					0.2	5			7.5		
4,4'-Methylenedianiline	101-77-9	118										0.081	0.81
Naphthalene	91-20-3	119									130		
Nickel acetate	373-02-4	186											
Nickel carbonyl	13463-39-3	186		^{f}	0.25^{f}	0.031^{f}					1.4		
Nickel chloride	7718-54-9	186										0.6	0.6
Nickel compounds	7440-02-0	186								0.006	1		
Nickel nitrate	13138-45-9	186										3	3
Nickel oxide	1313-99-1	186										0.75	0.75
Nickel refinery dust	NI_DUST	186											
Nickel subsulfide	12035-72-2	186											
Nickel sulfate	7786-81-4	186										2.5	2.5

AEGL = Acute exposure guideline levels for mild effects (AEGL-1) and moderate effects (AEGL-2) for 1- and 8-hour exposures. Superscripts indicate the AEGL's status: f = final, i=interim, and p=proposed. **ERPG** = US DOE Emergency Removal Program guidelines for mild or transient effects (ERPG-1) and irreversible or serious effects (ERPG-2) for 1-hour exposures. **MRL** = ATSDR minimum risk levels for no adverse effects for 1 to 14-day exposures. **REL** = California EPA reference exposure level for no adverse effects. Most, but not all, RELs are for 1-hour exposures. **IDLH/10** = One-tenth of levels determined by NIOSH to be imminently dangerous to life and health, approximately comparable to mild effects levels for 1-hour exposures. **TEEL** = US DOE Temporary emergency exposure limits for no effects (TEEL-0) and mild, transient effects (TEEL-1) for 1-hour exposures. *TEELs are derived according to a tiered, formula-like methodology, and do not undergo peer review. They are not recommended as the basis for regulatory decision-making and are shown here only to inform situations where acute values from other sources are not available.*

			AEGL-1 (1-h)	AEGL-1 (8-h)	AEGL-2 (1-h)	AEGL-2 (8-h)	ERPG-1	ERPG-2	MRL	REL	IDLH/10	TEEL-0	TEEL-1
CHEMICAL NAME	CAS NO.	HAP NO.	mg/m^3	mg/m^3	mg/m^3	mg/m^3	mg/m^3	mg/m^3	mg/m^3	mg/m^3	mg/m^3	mg/m^3	mg/m^3
Nitrobenzene	98-95-3	120									100		
4-Nitrobiphenyl	92-93-3	121										0.25	0.75
4-Nitrophenol	100-02-7	122										0.75	2.5
2-Nitropropane	79-46-9	123									36		
Nitrosodimethylamine	62-75-9	125										3.5	10
N-Nitrosomorpholine	59-89-2	126										12	30
N-Nitroso-N-methylurea	684-93-5	124										0.015	0.05
Parathion	56-38-2	127		[i]	1.5[i]	0.48[i]					1		
Polychlorinated biphenyls	1336-36-3	136										1	3
Aroclor 1016	12674-11-2	136										0.2	0.6
Aroclor 1221	11104-28-2	136										0.2	0.6
Aroclor 1242	53469-21-9	136										1	3
Aroclor 1248	12672-29-6	136										0.2	0.6
Aroclor 1254	11097-69-1	136										0.5	1.5
Aroclor 1260	11096-82-5	136										0.3	0.75
Pentachloronitrobenzene	82-68-8	128										0.5	1.5
Pentachlorophenol	87-86-5	129									0.25		
Phenol	108-95-2	130	58[f]	24[f]	89[f]	46[f]	38	190		5.8	96		
p-Phenylenediamine	106-50-3	131										0.1	0.3
Phosgene	75-44-5	132		[f]	1.2[f]	0.16[f]		2		0.004	0.81		
Phosphine	7803-51-2	133		[f]	2.8[f]	0.35[f]		0.7					
Phosphorus, white	7723-14-0	134							0.02				
Phthalic anhydride	85-44-9	135									6		
Acenaphthene	83-32-9	187										0.4	1.2

AEGL = Acute exposure guideline levels for mild effects (AEGL-1) and moderate effects (AEGL-2) for 1- and 8-hour exposures. Superscripts indicate the AEGL's status: f = final, i=interim, and p=proposed. **ERPG** = US DOE Emergency Removal Program guidelines for mild or transient effects (ERPG-1) and irreversible or serious effects (ERPG-2) for 1-hour exposures. **MRL** = ATSDR minimum risk levels for no adverse effects for 1 to 14-day exposures. **REL** = California EPA reference exposure level for no adverse effects. Most, but not all, RELs are for 1-hour exposures. **IDLH/10** = One-tenth of levels determined by NIOSH to be imminently dangerous to life and health, approximately comparable to mild effects levels for 1-hour exposures. **TEEL** = US DOE Temporary emergency exposure limits for no effects (TEEL-0) and mild, transient effects (TEEL-1) for 1-hour exposures. *TEELs are derived according to a tiered, formula-like methodology, and do not undergo peer review. They are not recommended as the basis for regulatory decision-making and are shown here only to inform situations where acute values from other sources are not available.*

			AEGL-1 (1-h)	AEGL-1 (8-h)	AEGL-2 (1-h)	AEGL-2 (8-h)	ERPG-1	ERPG-2	MRL	REL	IDLH/10	TEEL-0	TEEL-1
CHEMICAL NAME	CAS NO.	HAP NO.	mg/m^3	mg/m^3	mg/m^3	mg/m^3	mg/m^3	mg/m^3	mg/m^3	mg/m^3	mg/m^3	mg/m^3	mg/m^3
Anthracene	120-12-7	187										2	6
Benzo(a)anthracene	56-55-3	187										0.1	0.3
Benzo(b)fluoranthene	205-99-2	187										0.2	0.6
Benzo(k)fluoranthene	207-08-9	187										0.2	0.6
Benzo(g,h,i)perylene	191-24-2	187										10	30
Benzo(a)pyrene	50-32-8	187										0.2	0.6
Carbazole	86-74-8	187										0.75	2.5
beta-Chloronaphthalene	91-58-7	187										0.2	0.6
Chrysene	218-01-9	187										0.2	0.6
Dibenz(a,h)anthracene	53-70-3	187										10	30
Dibenzo[a,e]pyrene	192-65-4	187										0.035	0.1
Fluoranthene	206-44-0	187										0.005	0.015
Fluorene	86-73-7	187										7.5	25
Indeno(1,2,3-cd)pyrene	193-39-5	187										0.15	0.5
3-Methylcholanthrene	56-49-5	187										0.2	0.6
1-Methylnaphthalene	90-12-0	187										6	20
2-Methylnaphthalene	91-57-6	187										6	20
2-Naphthylamine	91-59-8	187										2.5	7.5
1-Nitropyrene	5522-43-0	187										0.1	0.3
Phenanthrene	85-01-8	187										0.4	1
Pyrene	129-00-0	187										15	15
1,3-Propane sultone	1120-71-4	137										0.4	1.2
beta-Propiolactone	57-57-8	138										1.5	1.5
Propionaldehyde	123-38-6	139	110^{i}	110^{i}	620^{i}	260^{i}							

AEGL = Acute exposure guideline levels for mild effects (AEGL-1) and moderate effects (AEGL-2) for 1- and 8-hour exposures. Superscripts indicate the AEGL's status: f = final, i=interim, and p=proposed. **ERPG** = US DOE Emergency Removal Program guidelines for mild or transient effects (ERPG-1) and irreversible or serious effects (ERPG-2) for 1-hour exposures. **MRL** = ATSDR minimum risk levels for no adverse effects for 1 to 14-day exposures. **REL** = California EPA reference exposure level for no adverse effects. Most, but not all, RELs are for 1-hour exposures. **IDLH/10** = One-tenth of levels determined by NIOSH to be imminently dangerous to life and health, approximately comparable to mild effects levels for 1-hour exposures. **TEEL** = US DOE Temporary emergency exposure limits for no effects (TEEL-0) and mild, transient effects (TEEL-1) for 1-hour exposures. *TEELs are derived according to a tiered, formula-like methodology, and do not undergo peer review. They are not recommended as the basis for regulatory decision-making and are shown here only to inform situations where acute values from other sources are not available.*

			AEGL-1 (1-h)	AEGL-1 (8-h)	AEGL-2 (1-h)	AEGL-2 (8-h)	ERPG-1	ERPG-2	MRL	REL	IDLH/10	TEEL-0	TEEL-1
CHEMICAL NAME	CAS NO.	HAP NO.	mg/m^3	mg/m^3	mg/m^3	mg/m^3	mg/m^3	mg/m^3	mg/m^3	mg/m^3	mg/m^3	mg/m^3	mg/m^3
Propoxur	114-26-1	140										0.5	1.5
Propylene dichloride	78-87-5	141							0.23		180		
Propylene oxide	75-56-9	142	170[f]	170[f]	690[f]	200[f]	120	590		3.1	95		
1,2-Propyleneimine	75-55-8	143		[f]	28[f]	2.8[f]							
Quinoline	91-22-5	144										1.1	3.2
Quinone	106-51-4	145									10		
Selenium compounds	7782-49-2	189									0.1		
Hydrogen selenide	7783-07-5	189		[i]	2.4[i]	0.86[i]		0.66		0.005	0.33		
Potassium selenite	7790-59-2	189										0.5	1.5
Selenious acid	7783-00-8	189										0.3	1
Selenium dioxide	7446-08-4	189										0.25	0.75
Selenium disulfide	7488-56-4	189										0.35	1
Selenium oxychloride	7791-23-3	189										0.4	1.2
Selenium sulfide	7446-34-6	189										0.25	0.75
Sodium selenite	13410-01-0	189										0.5	1.5
Sodium selenite	10102-18-8	189										0.4	1.2
Styrene	100-42-5	146	85[i]	85[i]	550[i]	550[i]	210	1100		21	300		
Styrene oxide	96-09-3	147										20	61
2,3,7,8-Tetrachlorodibenzo-p-dioxin	1746-01-6	148										0.0006	0.0015
1,1,2,2-Tetrachloroethane	79-34-5	149									69		
Tetrachloroethene	127-18-4	150	240[i]	240[i]	1600[i]	550[i]	680	1400	1.4	20	100		
Titanium tetrachloride	7550-45-0	151	0.54[i]	0.54[i]	7.8[i]	0.73[i]	5	20					
Toluene	108-88-3	152	750[i]	750[i]	4500[i]	2400[i]	190	1100	3.8	37	190		
2,4-Toluene diamine	95-80-7	153										4	12

AEGL = Acute exposure guideline levels for mild effects (AEGL-1) and moderate effects (AEGL-2) for 1- and 8-hour exposures. Superscripts indicate the AEGL's status: f = final, i=interim, and p=proposed. **ERPG** = US DOE Emergency Removal Program guidelines for mild or transient effects (ERPG-1) and irreversible or serious effects (ERPG-2) for 1-hour exposures. **MRL** = ATSDR minimum risk levels for no adverse effects for 1 to 14-day exposures. **REL** = California EPA reference exposure level for no adverse effects. Most, but not all, RELs are for 1-hour exposures. **IDLH/10** = One-tenth of levels determined by NIOSH to be imminently dangerous to life and health, approximately comparable to mild effects levels for 1-hour exposures. **TEEL** = US DOE Temporary emergency exposure limits for no effects (TEEL-0) and mild, transient effects (TEEL-1) for 1-hour exposures. *TEELs are derived according to a tiered, formula-like methodology, and do not undergo peer review. They are not recommended as the basis for regulatory decision-making and are shown here only to inform situations where acute values from other sources are not available.*

			AEGL-1 (1-h)	AEGL-1 (8-h)	AEGL-2 (1-h)	AEGL-2 (8-h)	ERPG-1	ERPG-2	MRL	REL	IDLH/10	TEEL-0	TEEL-1
CHEMICAL NAME	CAS NO.	HAP NO.	mg/m^3	mg/m^3	mg/m^3	mg/m^3	mg/m^3	mg/m^3	mg/m^3	mg/m^3	mg/m^3	mg/m^3	mg/m^3
2,4/2,6-Toluene diisocyanate mixture (TDI)	26471-62-5	154										1.8	5.3
2,4-Toluene diisocyanate	584-84-9	154	0.14^f	0.071^f	0.59^f	0.15^f	0.071	1.1			1.8		
o-Toluidine	95-53-4	155									22		
Toxaphene	8001-35-2	156										0.5	1
1,2,4-Trichlorobenzene	120-82-1	157										37	37
1,1,2-Trichloroethane	79-00-5	158									55		
Trichloroethylene	79-01-6	159	700^i	410^i	2400^i	1300^i	540	2700	11				
2,4,5-Trichlorophenol	95-95-4	160										10	30
2,4,6-Trichlorophenol	88-06-2	161										10	30
Triethylamine	121-44-8	162								2.8			
Trifluralin	1582-09-8	163										0.025	0.075
2,2,4-trimethylpentane	540-84-1	164										350	350
Uranium compounds	7440-61-1	188									1		
Uranium (IV) dioxide	1344-57-6	188						10					
Uranium hexafluoride	7783-81-5	188	3.6^f	^{f}	9.6^f	1.2^f	5	15					
Uranium oxide	1344-59-8	188						10					
Uranium, soluble salts	URANSOLS	188										0.05	0.6
Uranyl acetate dihydrate	541-09-3	188										0.075	1
Uranyl nitrate hexahydrate	13520-83-7	188										0.1	1.2
Vinyl acetate	108-05-4	165	24^i	24^i	630^i	260^i	18	260					
Vinyl bromide	593-60-2	166										22	66

AEGL = Acute exposure guideline levels for mild effects (AEGL-1) and moderate effects (AEGL-2) for 1- and 8-hour exposures. Superscripts indicate the AEGL's status: f = final, i=interim, and p=proposed. **ERPG** = US DOE Emergency Removal Program guidelines for mild or transient effects (ERPG-1) and irreversible or serious effects (ERPG-2) for 1-hour exposures. **MRL** = ATSDR minimum risk levels for no adverse effects for 1 to 14-day exposures. **REL** = California EPA reference exposure level for no adverse effects. Most, but not all, RELs are for 1-hour exposures. **IDLH/10** = One-tenth of levels determined by NIOSH to be imminently dangerous to life and health, approximately comparable to mild effects levels for 1-hour exposures. **TEEL** = US DOE Temporary emergency exposure limits for no effects (TEEL-0) and mild, transient effects (TEEL-1) for 1-hour exposures. *TEELs are derived according to a tiered, formula-like methodology, and do not undergo peer review. They are not recommended as the basis for regulatory decision-making and are shown here only to inform situations where acute values from other sources are not available.*

			AEGL-1 (1-h)	AEGL-1 (8-h)	AEGL-2 (1-h)	AEGL-2 (8-h)	ERPG-1	ERPG-2	MRL	REL	IDLH/10	TEEL-0	TEEL-1
CHEMICAL NAME	CAS NO.	HAP NO.	mg/m^3	mg/m^3	mg/m^3	mg/m^3	mg/m^3	mg/m^3	mg/m^3	mg/m^3	mg/m^3	mg/m^3	mg/m^3
Vinyl chloride	75-01-4	167	640^i	180^i	3100^i	2100^i	1300	13000	1.3	180			
Vinylidene chloride	75-35-4	168										20	79
m-Xylene	108-38-3	171								22	390		
o-Xylene	95-47-6	170								22	390		
p-Xylene	106-42-3	172								22	390		
Xylenes (mixed)	1330-20-7	169	560^i	560^i	4000^i	1700^i			8.7	22	390		

Source: USEPA

Appendix B

Typical Maintenance and Troubleshooting for Air Stripping Systems

Equipment Type	Process and Operational Monitoring Parameters	Typical Maintenance Items		Troubleshooting Approaches
		Activity	Maintenance Interval	
General Comment: In undertaking any installation, operation, and/or maintenance activity related to live electrical loads, stored mechanical or hydraulic energy, and/or hazardous chemicals, review and observe federal OSHA codes, guidelines, and best practices, in addition to any state or province-specific health and safety regulations for the industrial workplace.				
Tower	Tower Verticality and Plumbness	Check with level	1 year, and after major storm events for exterior installations	Vertical alignment tolerance is $\frac{1}{2}$ in. per 10 ft of tower height. Restore plumbness via shimming, seal w/ grout after shim install. For acute misalignment (i.e., imminent fall hazard), quarantine the fall zone, remove the tower with a crane, evaluate for structural integrity before reinstallation.
	Tower Coatings: Interior	Remove packing and tower mechanical internals, inspect interior coatings for integrity, spalling, holidays, and potential pitting if process water is chemically aggressive.	2 year	Consult coatings maintenance system and appropriate surface prep and coatings manuals of practice. For spot repair, clean and re-coat in-kind. For universal pitting or spalling, take tower out of service, consult with for manufacturer for recommendations for field repair or replacement.
	Tower Coatings: Exterior	Check coatings for spalling, chipping, and overall integrity	5 year	See coatings approach recommendations for Interior Tower Coatings. NSF-certification not required for exterior coatings
	Tower Anchor Lugs	Check nuts and lag bolts for thread binding, rust, material loss, and spalling if coated	5 year	For surficial spalling or coatings loss, brush-blast and re-coat. For deeper metal loss or thread binding, consider cutting anchor bolts and reinstalling with a mechanical or adhesive anchor. Consult a structural engineer.
	Tower Ladder and Platform Integrity	Check ladder connections and weld integrity, maintenance any safety systems such as but not limited to fall protection wire rope or safety harnesses.	1 year	Perform manufacturer-suggested maintenance on fall protection gear. As a life safety issue, utilities may wish to write a maintenance contract with a professional industrial hygienist or safety compliance contractor to perform regular equipment maintenance.

Equipment Type	Process and Operational Monitoring Parameters	Typical Maintenance Items		Troubleshooting Approaches
		Activity	Maintenance Interval	
General Comment: In undertaking any installation, operation, and/or maintenance activity related to live electrical loads, stored mechanical or hydraulic energy, and/or hazardous chemicals, review and observe federal OSHA codes, guidelines, and best practices, in addition to any state or province-specific health and safety regulations for the industrial workplace.				
	Tower Internals Integrity	Check tower internals for structural integrity, especially if process water is chemically aggressive	1 year	Replace deteriorating tower internals with chemically inert materials. Take consideration for site-specific aquatic chemistry. Consider stainless steel or aluminum internal fittings and distribution piping, fiberglass-reinforced plastic (FRP) packing trays.
Packing	Packing Fouling (Hardness or Metal Precipitates such as Fe or Mn)	Inspect packing at upper and lower regions in tower height, and at an interim location if available. Inspect to a packing depth of a minimum of 12 in. below upper surface of packing. At inspection ports if available, review packed bed differential data for upward trend. Compare to manufacturer recommendations for DP.	Quarterly, or at the beginning and end of each operating cycle if source is operated seasonally	Potential acid wash in-situ or evaluate process feasibility of an NSF-certified acid or sequesterant feed. Remove interior tower mechanicals, remove packing and acid wash in a controlled, contained environment, or replace media. After packing replacement, reinstall flanged tower ports with minimum $\frac{1}{4}$ in. closed-cell Teflon sponge gasket. Reinstall packing to within 6 in. +/– of the tower body flange.
	Biofilm Growth, or "Slime" Buildup	Inspect packing at inspection ports if available, review packed bed differential data for upward trend. Compare to manufacturer recommendations for DP.	Quarterly, or at the beginning and end of each operating cycle if source is operated seasonally	Potential shock disinfection to alleviate immediate problem. Possible cleaning of packing media in an oxidant bath (chlorine, hydrogen peroxide). Consider pre-tower oxidant feed to mitigate biofilm growth.

Equipment Type	Process and Operational Monitoring Parameters	Typical Maintenance Items		Troubleshooting Approaches
		Activity	Maintenance Interval	
General Comment: In undertaking any installation, operation, and/or maintenance activity related to live electrical loads, stored mechanical or hydraulic energy, and/or hazardous chemicals, review and observe federal OSHA codes, guidelines, and best practices, in addition to any state or province-specific health and safety regulations for the industrial workplace.				
	Blower Current	Take regular measurements of blower current as a proxy value for determining increase in fouling	Monthly	Most constant-speed motors consume less horsepower (at reduced current) when operated against increased resistance to airflow. Read current using a calibrated digital ammeter, accompanied by voltage readings on all three legs (assuming typical 3-phase, 480 V operation of majority of blowers).
	Packing "Crush"	Inspect packing to ensure structural integrity remains intact.	1 year, and after initial installation and/or any packing removal and reinstall for maintenance	Install more interim packing trays. Use a stronger packing model with similar Henry's law characteristics to basis-of-design packing.
Diffuser or Bubbler	Structural Integrity	Inspect diffuser header(s) and bubbler arms for structural integrity, pipe thinning, and pitting.	1 year	Replace failing components upon detection
	Outlet Port Blinding	Inspect bubble arm outlets and/or nozzles for blinding by hardness and/or precipitated metals	1 year, or more frequently for process water with a tendency to precipitate solids	Remove and clean/overhaul distribution arms with hand tools. Consider a weak acid bath for routine cleanings and/or physical cleaning and drilling out clogged outlets.
	Bubble size distribution	If the bubbles are too small, that is an indication of clogging on the diffuser port If the bubbles are too large, the port has eroded out.	Monthly	Replace failing components upon detection

Equipment Type	Process and Operational Monitoring Parameters	Typical Maintenance Items		Troubleshooting Approaches
		Activity	Maintenance Interval	
General Comment: In undertaking any installation, operation, and/or maintenance activity related to live electrical loads, stored mechanical or hydraulic energy, and/or hazardous chemicals, review and observe federal OSHA codes, guidelines, and best practices, in addition to any state or province-specific health and safety regulations for the industrial workplace.				
Tray	Structural Integrity	Check tray structural integrity and deflection, especially at center of tray, especially in systems which tend to scale the packing with hardness or metals	1 year	If cracking is visible, remove and replace tray. If scale has built up, consider an acid feed or upsize the gauge of the tray grid.
	Flow distribution	Check tray for mineral scale blinding off tray holes	1 year	Remove and clean/overhaul trays. Cleaning may require a weak acid bath to assist in removing deposits.
Spray Laterals and Nozzles	Structural Integrity	Inspect laterals and headers for structural integrity and bending.	1 year	Replace failing components upon detection.
	Flow distribution	Check that water output is even and consistent. Uneven or inconsistent spray is indication of a clog.	1 year	Remove and clean nozzles.
Venturi	See spray laterals and nozzles			

Equipment Type	Process and Operational Monitoring Parameters	Typical Maintenance Items		Troubleshooting Approaches
		Activity	Maintenance Interval	
General Comment: In undertaking any installation, operation, and/or maintenance activity related to live electrical loads, stored mechanical or hydraulic energy, and/or hazardous chemicals, review and observe federal OSHA codes, guidelines, and best practices, in addition to any state or province-specific health and safety regulations for the industrial workplace.				
Blower or Compressor, and Ductwork	Flex Bellows Integrity	Check for cracking and flexibility. Bellows should be installed on intake (if ductwork is used) and output of blower.	1 year	Replace if physical integrity is compromised.
	Blower Motor Guard and/or Weather Enclosure	Ensure correct mounting	1 year	First, ensure the motor power supply is disconnected before performing any close inspection or maintenance. If loose, replace and reinstall bolts. Ensure OSHA-compliant lock-out/tag-out equipment is installed.
	Belt Check	Verify belt tension	After first 24 hours of operation, then Quarterly, or at the beginning and end of each operating cycle if source is operated seasonally	Perform belt tensioning per manufacturer's recommendation.
	Blower and Motor Bearings	Check lubricant level and top-off grease	1 year, or at the beginning and end of each operating cycle if source is operated seasonally. 6 months if dirty environment.	If true binding is ever observed, open case and inspect bearings. If structurally sound, reinstall casing and perform a bearing system overhaul—discharge a full cavity-load of bearing grease and pump in new grease to the bearing casing. If bearings are compromised structurally, call manufacturer's representative for service visit. Vibration analysis and mitigation should be performed by a qualified/experienced technician.

Equipment Type	Process and Operational Monitoring Parameters	Typical Maintenance Items		Troubleshooting Approaches
		Activity	Maintenance Interval	
General Comment: In undertaking any installation, operation, and/or maintenance activity related to live electrical loads, stored mechanical or hydraulic energy, and/or hazardous chemicals, review and observe federal OSHA codes, guidelines, and best practices, in addition to any state or province-specific health and safety regulations for the industrial workplace.				
	Motor Maintenance	Check for dirt and dust accumulation at vents, examine for signs of moisture, unusual noise, oil leak at seals, vibration, integrity of foundation pad. Checking running current, insulation and winding resistance, and other electrical parameters.	6 months	Maintenance personnel should perform regular cleaning and re-fill of motor bearing lubricant. At detection of improper vibration or other major equipment failure, consult a manufacturer's service representative.
	Motor Electrical Characteristics	Maintenance record of line voltage, amp draw, and blower RPM.	Monthly	Use record as reference in the event of severe operational problem or failure.
	Fan Maintenance	Ensure fan blades are clean and balanced, with no unusual vibration.	6 months	Maintenance personnel should perform regular cleaning and re-fill of motor bearing lubricant. At detection of improper vibration or other major equipment failure, consult a manufacturer's service representative.
	Bearing Locking Collars	Check collars for correct grip on drive shaft, verify no signs of slippage	6 months	Tighten locking collar and/or set screw to manufacturer recommended setting torque.
	Ductwork Integrity	Check for leaks	1 year	Repair leak. If required, replace ductwork.

Equipment Type	Process and Operational Monitoring Parameters	Typical Maintenance Items		Troubleshooting Approaches
		Activity	Maintenance Interval	
General Comment: In undertaking any installation, operation, and/or maintenance activity related to live electrical loads, stored mechanical or hydraulic energy, and/or hazardous chemicals, review and observe federal OSHA codes, guidelines, and best practices, in addition to any state or province-specific health and safety regulations for the industrial workplace.				
Air Intake	Screen Inspection	Check screen and filter cartridge for debris (e.g., leaves), dirt, and dust buildup.	Monthly, and after major storms.	Replace filter cartridge.
	Louver Inspection	If static, verify cleanliness of louver. If motorized, verify cleanliness of louver blades and check louver motor operator in accordance with motor maintenance itinerary reviewed above.	6 months	
Process Instrumentation	Blower Pressure Switch Condition		Per manufacturer's recommendations	
	Temperature switch condition		Per manufacturer's recommendations	
	Packed Bed Differential Pressure Gauge Condition		Per manufacturer's recommendations	

Index

Note: *f.* indicates figure; *t.* indicates table

D

E

F

G

V

W

This page intentionally blank.

AWWA Manuals

M1 *Principles of Water Rates, Fees, and Charges,* #30001
M2 *Instrumentation and Control,* #30002
M3 *Safety Management for Water Utilities,* #30003
M4 *Water Fluoridation Principles and Practices,* #30004
M5 *Water Utility Management,* #30005
M6 *Water Meters—Selection, Installation, Testing, and Maintenance,* #30006
M7 *Problem Organisms in Water: Identification and Treatment,* #30007
M9 *Concrete Pressure Pipe,* #30009
M11 *Steel Pipe—A Guide for Design and Installation,* #30011
M12 *Simplified Procedures for Water Examination,* #30012
M14 *Backflow Prevention and Cross-Connection Control: Recommended Practices,* #30014
M17 *Fire Hydrants: Installation, Field Testing, and Maintenance,* #30017
M19 *Emergency Planning for Water and Wastewater Utilities,* #30019
M20 *Water Chlorination/Chloramination Practices and Principles,* #30020
M21 *Groundwater,* #30021
M22 *Sizing Water Service Lines and Meters,* #30022
M23 *PVC Pipe—Design and Installation,* #30023
M24 *Planning for the Distribution of Reclaimed Water,* #30024
M25 *Flexible-Membrane Covers and Linings for Potable-Water Reservoirs,* #30025
M27 *External Corrosion Control for Infrastructure Sustainability,* #30027
M28 *Rehabilitation of Water Mains,* #30028
M29 *Water Utility Capital Financing,* #30029
M30 *Precoat Filtration,* #30030
M31 *Distribution System Requirements for Fire Protection,* #30031
M32 *Computer Modeling of Water Distribution Systems,* #30032
M33 *Flowmeters in Water Supply,* #30033
M36 *Water Audits and Loss Control Programs,* #30036
M37 *Operational Control of Coagulation and Filtration Processes,* #30037
M38 *Electrodialysis and Electrodialysis Reversal,* #30038
M41 *Ductile-Iron Pipe and Fittings,* #30041
M42 *Steel Water-Storage Tanks,* #30042
M44 *Distribution Valves: Selection, Installation, Field Testing, and Maintenance,* #30044
M45 *Fiberglass Pipe Design,* #30045
M46 *Reverse Osmosis and Nanofiltration,* #30046
M47 *Capital Project Delivery,* #30047
M48 *Waterborne Pathogens,* #30048
M49 *Quarter-Turn Valves: Head Loss, Torque, and Cavitation Analysis,* #30049
M50 *Water Resources Planning,* #30050
M51 *Air Valves: Air-Release, Air/Vacuum, and Combination,* #30051
M52 *Water Conservation Programs: A Planning Manual,* #30052
M53 *Microfiltration and Ultrafiltration Membranes for Drinking Water,* #30053
M54 *Developing Rates for Small Systems,* #30054
M55 *PE Pipe—Design and Installation,* #30055
M56 *Nitrification Prevention and Control in Drinking Water,* #30056
M57 *Algae: Source to Treatment,* #30057
M58 *Internal Corrosion Control in Water Distribution Systems,* #30058
M60 *Drought Preparedness and Response,* #30060
M61 *Desalination of Seawater,* #30061
M62 *Membrane Applications for Water Reuse,* #30062
M63 *Aquifer Storage and Recovery,* #30063

M65 *On-Site Generation of Hypochlorite,* #30065

M66 *Cylinder and Vane Actuators and Controls—Design and Installation,* #30066

M68 *Water Quality in Distribution Systems,* #30068

M69 *Inland Desalination and Concentrate Management,* #30069

M71 *Climate Action Plans—Adaptive Management Strategies for Utilities*

M72 *Knife Gate Valves*

M75 *Elastomers for Waterworks: Pipes, Valves, and Fittings,* #30075

M77 *Condition Assessment of Water Mains,* #30077

www.ingramcontent.com/pod-product-compliance
Lightning Source LLC
LaVergne TN
LVHW082342200926
848481LV00012B/3424
* 9 7 8 1 6 4 7 1 7 0 2 6 4 *